AF832889

QUANTUM OPTICS AND LASER EXPERIMENTS

Edited by **Sergiy Lyagushyn**

Quantum Optics and Laser Experiments
Edited by Sergiy Lyagushyn

Published by InTech

Janeza Trdine 9, 51000 Rijeka, Croatia

Copyright © 2011 InTech

All chapters are distributed under the Creative Commons Attribution 3.0 license, which allows users to download, copy and build upon published articles even for commercial purposes, as long as the author and publisher are properly credited, which ensures maximum dissemination and a wider impact of our publications. After this work has been published by InTech, authors have the right to republish it, in whole or part, in any publication of which they are the author, and to make other personal use of the work. Any republication, referencing or personal use of the work must explicitly identify the original source.

As for readers, this license allows users to download, copy and build upon published chapters even for commercial purposes, as long as the author and publisher are properly credited, which ensures maximum dissemination and a wider impact of our publications.

Notice

Statements and opinions expressed in the chapters are these of the individual contributors and not necessarily those of the editors or publisher. No responsibility is accepted for the accuracy of information contained in the published chapters. The publisher assumes no responsibility for any damage or injury to persons or property arising out of the use of any materials, instructions, methods or ideas contained in the book.

Publishing Process Manager Marina Jozipovic

Technical Editor Teodora Smiljanic

Cover Designer InTech Design Team

Additional hard copies can be obtained from orders@intechopen.com

Quantum Optics and Laser Experiments, Edited by Sergiy Lyagushyn
p. cm.
ISBN 978-953-307-937-0

We are IntechOpen, the world's leading publisher of Open Access books

Built by scientists, for scientists

4,100+
Open access books available

116,000+
International authors and editors

120M+
Downloads

151
Countries delivered to

Our authors are among the
Top 1%
most cited scientists

12.2%
Contributors from top 500 universities

CLARIVATE ANALYTICS BOOK CITATION INDEX INDEXED

WEB OF SCIENCE™

Selection of our books indexed in the Book Citation Index
in Web of Science™ Core Collection (BKCI)

Interested in publishing with us?
Contact book.department@intechopen.com

Numbers displayed above are based on latest data collected.
For more information visit www.intechopen.com

For EU product safety concerns:
IN TECH d.o.o., Prolaz Marije Krucifikse Kozulić 3, 51000 Rijeka, Croatia,
info@intechopen.com or visit our website at intechopen.com.

Meet the editor

Dr Sergiy F. Lyagushyn was born in 1955 in Dnipropetrovs'k (Ukraine, USSR). He graduated from the Physical Department of the M.V. Lomonosov Moscow State University in 1978. He studied and worked at the Quantum Statistics and Field Theory Division headed by Academician N.N. Bogolyubov. He finished his post-graduate courses in Moscow State University in 1988 (supervisor – Prof. N.N. Bogolyubov (Jr.)). He has worked in Dnipropetrovs'k since 1988. He dealt with problems of rocket thrusters and space power supply in Experimental Design Office "Foton". He published a number of papers devoted to renewable power sources, diffraction optical elements, and space solar power. Dr Lyagushyn has served as a member of two committees of the International Astronautical Federation in 1990s. Since 1992 he has been an Associate Professor at the Quantum Macrophysics Division of Dnipropetrovs'k University (now – Theoretical Physics Division), and from 1994-2007 was the Deputy Dean of the Physics Department. His present sphere of scientific interests includes nonequilibrium statistical physics and quantum optics. His main publications in this field were in Physica A, Physics of Particles and Nuclei, Journal of Dnipropetrovs'k National University (Physics & Radioelectronics series).

Contents

Preface

The book covers a wide spectrum of research problems concerning quantum theory of light and experiments using its quantum properties. In reference literature one can find a number of definitions for the term "quantum optics" – from the sphere of phenomena revealing the quantum nature of light to the optics section dealing with statistical properties of emission. Such a contradictory situation reflects a complicated way of notion formation. Difficulties of the classical wave concept of light were the basis for formulating new quantum concepts of light emission, propagation, interaction and a new general concept – field concept of matter in the first quarter of the 20th century. In this sense we can speak about quantum optics whenever optical phenomena are considered from the position of quantum theory. That is especially true for the world of phenomena that can be discussed and understood only within the framework of quantum picture. The possibilities of optical experiments have been broadened fantastically by the invention of laser. Since lasers are quantum optical generators, the domain of experiments with laser emission seems to be close to quantum optics. At the same time, strong electromagnetic fields generated by lasers usually manifest their classical properties, so some analysis is necessary for including the observed phenomena into the field of quantum optics. In classical optics, correlation properties of light connected with the statistical nature of a real experiment were discussed in terms of the conception "coherence". In the experiments of Hanbury Brown–Twiss with quantum detecting of light, the process of receiving an electromagnetic emission was considered a usual random process for the first time. Later on, the whole ideology of probability theory and stochastic processes was applied to optical phenomena using quantum detectors for analyzing the statistical (correlation) properties of electromagnetic fields in optics. It gives substantiation for quantum optics identification as a statistical theory of light.

The editor was faced with different interpretations of quantum optics while analyzing the chapter proposals. Upon examining them, he reached the conclusion that all theoretical and experimental papers were welcome if they contributed to our understanding of light as a quantum phenomenon. The accepted approach was to speak about quantum optics in a wide sense, i.e. different phenomena demonstrating the quantum nature of light together with theoretical constructs applied to them; and in the narrow sense, i.e. the statistical theory of light processes and its incarnation with quantum detecting schemes. The present title of the book reflects its real contents and breadth of topics.

The first section titled "Theoretical Fundamentals: Problem of Observables" includes four chapters united in the search for adequate mathematical apparatus for quantum electromagnetic field state description, taking into account experimental research possibilities. The first chapter "Description of Field States with Correlation Functions & Measurements in Quantum Optics" by Dr. S. Lyagushyn and Prof. A. Sokolovsky incorporates the discussion of basic approaches to field investigation in quantum optics. Since measurements with quantum detectors lead to Glauber correlation function and the Glauber-Sudarshan *P*-function is the most consumable tool for practical field description, such functions are regarded as an optimal way for field diagnostics. Then the Bogolyubov reduced description is constructed for a medium consisting of two-level emitters (the Dicke superfluorescence phenomenon) and plasma-field system. In such way, the connection is made using simultaneous correlation functions of field amplitudes for system evolution description, constructing differential equations for them, and coming to a quasiequilibrium statistical operator for system constituents at large times, the statistical operator permitting correlation function calculation. The necessity of considering binary correlation of field is substantiated. Such kinds of electrodynamics in media imply obtaining certain material equations. Various forms of correlation description are presented: one-particle density matrix, Wigner distribution function, and correlation functions of Glauber type. Correlation functions in the theory of radiation transfer and corresponding equations are considered. A way to field evolution description on the basis of a generating functional and Glauber-Sudarshan distribution connected with it is proposed.

The second chapter is "Nonclassical Features of Superpositions of Coherent and Squeezed States for Electromagnetic Fields in Time-Varying Media" by Prof. Choi Jeong Ryeol. The author is interested in light behavior in media with varying characteristic parameters, the situation promising several interesting applications. A special method of field quantization based on the invariant operator theory is used. Thus deriving quantum solutions for time-dependent Hamiltonians becomes possible. The exact wave functions for the system with time-varying parameters can be derived in Fock, coherent, and squeezed states. Then superpositions of quantum states are considered in the search for nonclassical properties (high-order squeezing, subpoissonian photon statistics, and oscillations in the photon-number distribution). Such analysis is based on the Wigner distribution function, allowing us to know the phase space distribution connected to a simultaneous measurement of position and momentum. The Wigner distribution function is regarded as quasiprobability distribution function and is widely used in explaining intrinsic quantum features that have no classical analogue.

The third chapter "Photon Localization Revisited" by Prof. P. Saari is devoted to the intriguing problem that is traditionally under discussion in literature on quantum optics. Such considerations are urgent for needs of near-field optics, cavity quantum electrodynamics, and quantum computing. Some new optical phenomena connected with photon localization have drawn scientists' interest in recent years (see, for example Chapter 5). In textbooks the problems of position wave function and measurable quantities locality are compared. The author presents a qualified review together with his original results. Restrictions on photon localization set by the Paley-Wiener theorem and their seeming violation for certain two-dimensional wave packets are discussed.

The fourth chapter "Fusion Frames and Dynamics of Open Quantum Systems" by Prof. A. Jamiołkowski deals with the problems of quantum tomography, which is a procedure for reconstructing properties of a quantum object on the basis of experimentally accessible data. The state reconstruction requires identifying the quorum of observables, providing a possibility to determine expectation values of physical quantities for which no measuring apparatuses are available. The problem is discussed in terms of a set of density operators on the Hilbert space of the quantum system states. The main purpose of this contribution is to discuss properties of some Krylov subspaces in a given Hilbert space as natural examples of fusion frames and their applications in reconstructing open quantum system trajectories.

The second section of the book under the title "Quantum Phenomena with Laser Radiation" includes three chapters. The fifth chapter "Quantum Optics Phenomena in Synthetic Opal Photonic Crystals" by Prof. V. Moiseyenko and Dr. M. Dergachov opens the section. It contains a useful review of optical phenomena in materials with a space modulation of dielectric constant at distances close to the light wavelengths (so called photonic band-gap structures or photonic crystals). Gaps in their photonic band structure represent frequency regions where electromagnetic waves are forbidden, irrespective of the spatial propagation directions. Since the photon density of states is equal to zero inside the band gaps, emission of light sources embedded in these crystals should be inhibited in these spectral regions. Besides the emission inhibition effect, a number of new optical phenomena in 3D photonic crystals, interesting from the applied point of view, are under intensive study now. The main research directions are the following: effects of light localization, radiation of photonic crystals filled with organic and inorganic luminophores near the edges of photonic band-gaps, radiation of quantum dots in photonic crystal volume, quantum optics phenomena in nano-structured materials based on photonic crystals and nonlinear optical substances, effects of the radiation field amplification in photonic crystals, increase of solar cells efficiency with the use of photonic crystals. Synthetic opal photonic crystals containing nonlinear optical substances give a good chance to observe quantum optics phenomena in spatially nonuniform media where the photon mean free path is close to the light wavelength. The following quantum optics phenomena are considered: luminescence, Raman scattering, and spontaneous parametric down-conversion. Experimental samples were made of nanodisperse globules of silica dioxide synthesized by authors. Results of spectral investigations are presented and discussed.

The sixth chapter "Resonant Effects of Quantum Electrodynamics in the Pulsed Light Field" by Prof. S. Roshchupkin et al. describes the great achievements of this group in investigating laser field influence on kinematics and cross-sections of various quantum electrodynamics processes of the both first and second orders in the fine structure constant, such as resonant spontaneous bremstrahlung of an electron scattered by a nucleus, resonant photocreation of electron-positron pair on a nucleus, resonant scattering of a lepton by a lepton, and resonant scattering of a photon by an electron in the field of a pulsed light wave. This scientific direction has been of great interest for many years. Theoretical study of such processes is based on solutions of the Dirac's equation for an electron in the field of a plane electromagnetic wave. Resonant character of strong field influence has common features with resonant interactions in quantum optics. The wide research activities presented in the chapter have resulted in some conclusions to be tested in experiments with accelerators in presence of strong fields.

The seventh chapter "Cold Atoms Experiments: Influence of Laser Intensity Imbalance on Cloud Formation" by Dr. I. Olivares and Dr. F. Aguilar can be regarded as an example showing the possibilities of modern laser experiments. The authors deal with a magneto optical trap with the intent to obtain a cloud of cold atoms. They describe an experiment that proved the stability of the cloud and the optical method to vary the laser intensity of the pump and trap beams. The influence of laser intensity imbalance on cloud formation is investigated and values for the threshold intensity of lasers supporting cloud formation are obtained. The technique of saturated absorption spectroscopy is described. The theoretical analysis is performed in terms of level populations and optical Bloch equations that is conventional for quantum optics.

The book's "geography" – from Estonia to Chile – shows the wide interest for the problems under discussion all over the world. The relative majority of authors from Ukraine reflect both the history of monograph formation and great potential of Ukrainian physics.

I am grateful to InTech's publishing team and especially to Publishing Process Manager Ms. Marina Jozipovic for their constructive approach to the book formation, understanding the authors' problems, and a tolerant attitude for my delays. I would like to thank my older colleagues Prof. A. Sokolovsky and Prof. V. Skalozub for their support at different stages of the Project.

Sergiy Lyagushyn,
Associate Professor of Theoretical Physics
Department of Oles' Honchar Dnipropetrovs'k National University
Ukraine

Part 1

Theoretical Fundamentals: Problem of Observables

1

Description of Field States with Correlation Functions and Measurements in Quantum Optics

Sergiy Lyagushyn and Alexander Sokolovsky
Oles' Honchar Dnipropetrovs'k National University
Ukraine

1. Introduction

Modern physics deals with the consistent quantum concept of electromagnetic field. Creation and annihilation operators allow describing pure quantum states of the field as excited states of the vacuum one. The scale of its changes obliges to use statistical description of the field. Therefore the main object for full description of the field is a statistical operator (density matrix). Field evolution is reflected by operator equations. If the evolution equations are formulated in terms of field strength operators, their general structure coincides with the Maxwell equations. At the same time from the point of view of experiments only reduced description of electromagnetic fields is possible. In order to analyze certain physical situations and use numerical methods, we have the necessity of passing to observable quantities that can be measured in experiments. The problem of parameters, which are necessary for non-equilibrium electromagnetic field description, is a key one for building the field kinetics whenever it is under discussion. The field kinetics embraces a number of physical theories such as electrodynamics of continuous media, radiation transfer theory, magnetic hydrodynamics, and quantum optics. In all the cases it is necessary to choose physical quantities providing an adequate picture of non-equilibrium processes after transfer to averages. It has been shown that the minimal set of parameters to be taken into account in evolution equations included binary correlations of the field. The corresponding theory can be built in terms of one-particle density matrices, Wigner distribution functions, and conventional simultaneous correlation functions of field operators. Obviously, the choice depends on traditions and visibility of phenomenon description. Some methods can be connected due to relatively simple relations expressing their key quantities through one another. The famous Glauber's analysis (Glauber, 1966) of a quantum detector operation had resulted in using correlation functions including positive- and negative-frequency parts of field operator amplitudes in the quantum optics field. Herewith the most interesting properties of field states are described with non-simultaneous correlation functions. Various approaches in theoretical and experimental research into field correlations are compared in the present chapter.

Our starting point is investigation of the Dicke superfluorescence (Dicke, 1954) on the basis of the Bogolyubov reduced description method (Akhiezer & Peletminskii, 1981). It paves the way to constructing the field correlation functions. We can give a relaxation process picture in different orders of the perturbation theory. The set of correlation functions providing a rather full description of the superfluorescence phenomenon obeys the set of differential equations. The further research into the correlation properties of the radiated field requires establishing the connection with the behavior of Glauber functions of different orders.

2. Electromagnetic field as an object of quantum statistical theory

A statistical operator ρ of electromagnetic field should take into account the whole variety of field modes and statistical structure abundance for each of them. Proceeding from the calculation convenience provided by using coherent states $|z\rangle$ of field modes, the Glauber-Sudarshan representation for the statistical operator of field (Klauder & Sudarshan, 1968) footholds in physics. We refer to the following view of this diagonal representation

$$\rho = \int d^2 z P\left(z, z^*\right) | z\rangle\langle z | \tag{1}$$

where $P(z,z^*)$ is so called P-distribution ($z \equiv \{z_{\alpha k}\}$ and these variables are numbered by polarization α and wave vector k of the field modes). Since coherent states form an overcrowded basis in the state space of the mode with the completeness condition

$$\frac{1}{\pi}\int d^2 z \,| z\rangle\langle z | = \hat{1}\,, \tag{2}$$

the most general representation for the statistical operator should include not only projection operators $| z\rangle\langle z |$, but also more general operator products $| z\rangle\langle z' |$. Nevertheless it can be shown (Glauber, 1969; Kilin, 2003) that a P-distribution can be obtained as a two-dimensional Fourier transformation of the generating functional

$$F\left(u, u^*\right) = \mathrm{Sp}\rho e^{\sum\limits_{\alpha k} u_{\alpha k} c_{\alpha k}^{+}} e^{-\sum\limits_{\alpha k} u_{\alpha k}^{*} c_{\alpha k}} \tag{3}$$

which is a generating one for all normally ordered field moments and can be calculated directly with an arbitrary statistical operator ρ. Here we use standard notation of quantum electrodynamics: $c_{\alpha k}^{+}$, $c_{\alpha k}$ are Bose amplitudes (creation and annihilation operators) of the field.

So we can use the representation (1) in all cases when the Fourier integral for (3) exists. Such situation embraces a great variety of states that are interesting for physicists. More general cases reveal themselves in singularities of the P-distribution, the representation (1) still being prospective for using if the P-distribution can be expressed via generalized functions of slow growth, i.e. δ-function and its derivatives. The term " P-distribution" is relatively conventional: function $P(z,z^*)$ is a real but non-positive one. Nevertheless, the field state description with the Glauber-Sudarshan P-distribution remains the most demonstrative and consumable. For example, a proposed definition of non-classical states of electromagnetic field (Bogolyubov (Jr.) et al., 1988) uses the expression (1) for the statistical operator. A state is referred to as non-classical one if one of two requirements is obeyed: either average number of photons in a mode is less than 1, or P-function is not positively determined or has singularity that is higher than the δ-function.

For a multi-mode field the statistical operator takes the form of a direct product of one-mode statistical operators. In Schrödinger picture the Liouville equation

$$\partial_t \rho(t) = -\frac{i}{\hbar}[\hat{H}, \rho(t)] \tag{4}$$

describes the evolution of an arbitrary physical system. In the case when electromagnetic field interacting with matter is under consideration the problem is reduced to the correct account of the matter influence, so some kinds of effective Hamiltonians may appear in an analogue of (4) for the statistical operator of field. Evolution description in Heisenberg picture seems to be closer to the classical one. We come to operator Maxwell equations for field operators with terms corresponding to the matter influence and demanding some kind of material equations.

More graphic way to describing the electromagnetic field, its states, and their evolution is using correlation functions of different types, i.e. averaged values of physical quantities characterizing the field. The problem of choosing them will be discussed below.

3. Correlation functions provided by methods of quantum optics

Conventional classical optics was very restricted in measuring the parameters of fields. All conclusions about properties of light including its polarization properties were drawn from measurements of light intensity, i.e. from values of some quadratic functions of the field (Landau & Lifshitz, 1988). Naturally, we speak now about transversal waves in vacuum. Regarding a wave, close to a monochromatic one, we use slowly varying complex amplitude $E_{0n}(t)$ for its description:

$$E_n = E_{0n}(t) e^{-i\omega t} . \tag{5}$$

Partially polarized light is characterized with the tensor of polarization

$$J_{mn} = \overline{E_{0m} E_{0n}^*} \tag{6}$$

where m and n corresponds to two possible directions of polarization and quick oscillations of field are neglected. Averaging is performed over time intervals or (in the case of statistically stable situation) in terms of probabilities. A sum of diagonal components of J_{mn} is a real value that is proportional to the field intensity (the energy flux density in the wave in our case). Note that the discussion of field correlation functions by Landau in the earlier edition of the mentioned book was one of the first in the literature.

A rather full analysis of the classical measurement picture is given in (Klauder & Sudarshan, 1968). It should be mentioned that real field parameters are obtained from complex conjugated values in this approach. Transition to the quantum electromagnetic theory (Scully & Zubairy, 1997) is connected with substitution of operator structures with creation and annihilation operators instead of complex conjugated functions and coming to positive- and negative-frequency parts of field operators. Such expressions will be shown later on.

Physical picture of field parameter registration in the quantum case can be reduced to the problem of photon detection. An ideal detector should have response that is independent of radiation frequency and be small enough in comparison with the scale of field changes. Generally accepted analysis of quantum photon detector (Glauber, 1965; Kilin, 2003) is based on using an atom in this role and regarding the operator of field-atom interaction in the electric dipole approximation

$$\hat{V} = -\hat{p}_n \cdot \hat{E}_n(x)$$

with $\hat{p}_n$ standing for the operator of the electric dipole moment of an atom localized in a point with a radius-vector x (we shall denote in such a simple way a three-dimensional spatial vector). The quantum theory derives the total probability w of atom transition from a definite initial ground state $|\mathrm{g}\rangle$ to an arbitrary final excited one $|\mathrm{e}\rangle$ belonging to the continuous spectrum during the time interval from t_0 to t on the basis of Dirac's nonstationary perturbation theory in the interaction picture (Kilin, 2003)

$$w = \int_{t_0}^{t} d\tau \int_{t_0}^{t} d\tau' \sum_{mn} R_{mn}(\tau - \tau') G_{mn}^{(1,1)}(x, \tau'; x, \tau) \tag{7}$$

where $R_{mn}(\tau - \tau')$ is a function of detector sensitivity and

$$G_{mn}^{(1,1)}(x_1, t_1; x_1', t_1') \equiv \langle \hat{E}_m^{(-)}(x_1, t_1) \hat{E}_n^{(+)}(x_1', t_1') \rangle \tag{8}$$

is field correlation function of the first order (we use the notation $\langle \hat{A} \rangle \equiv \mathrm{Sp}\, \rho \hat{A}$ for an arbitrary operator $\hat{A}$). Here and further we use standard expressions for operators of the vector potential, electric and magnetic field in the Coulomb gauge (Akhiezer A. & Berestetsky V., 1969)

$$\hat{A}_n(x) = c \sum_{k\alpha} \left(\frac{2\pi\hbar}{\omega_k V} \right)^{1/2} e_{\alpha k n} (c_{\alpha k} + c_{\alpha,-k}^{+}) e^{ikx} ; \tag{9}$$

$$\hat{E}_n(x) = i \sum_{k\alpha} \frac{(2\pi\hbar\omega_k)^{1/2}}{V^{1/2}} e_{\alpha k n} (c_{\alpha k} - c_{\alpha,-k}^{+}) e^{ikx} , \tag{10}$$

$$\hat{B}_n(x) = i \sum_{k\alpha} \frac{(2\pi\hbar\omega_k)^{1/2}}{V^{1/2}} \varepsilon_{nlm} \tilde{k}_l e_{\alpha k m} (c_{\alpha k} + c_{\alpha,-k}^{+}) e^{ikx}$$

In these formulas $e_{\alpha k n}$ are vectors of the circular polarization ($e_{\alpha k n} k_n = 0$), $\tilde{k}_l \equiv k_l / k$, $\omega_k = ck$, V is field volume. Field operators in (8) are the positive- and negative-frequency parts of electric field operator in the picture of interaction

$$\hat{E}_n(x,t) = \hat{E}_n^{(+)}(x,t) + \hat{E}_n^{(-)}(x,t) , \tag{11}$$

$$\hat{E}_n^{(+)}(x,t) \equiv i\sum_{k\alpha} \frac{(2\pi\hbar\omega_k)^{1/2}}{V^{1/2}} e_{\alpha kn} c_{\alpha k} e^{i(kx-\omega_k t)}, \qquad \hat{E}_n^{(-)}(x,t) = \hat{E}_n^{(+)}(x,t)^+ .$$

The correlation function of detector sensitivity in the suggestion that matrix elements of the dipole moment operator between the ground and excited states (so called dipole moment of transition) $\langle e | \hat{p}_n | g \rangle \equiv p_n$ are independent of a final state takes the form

$$R_{mn}(\tau - \tau') = \pi \frac{\nu}{\hbar^2} \overset{*}{p}_m p_n \delta(\tau - \tau') \equiv s_{mn} \delta(\tau - \tau') \tag{12}$$

where ν stands for the spectral density of states in the continuous spectrum. It is expedient to notice that the dependence of matrix elements of electric dipole moment on time in the interaction picture results in positive- and negative-frequency parts of field operators appearing in calculated averages.

It follows from (7) and (12) that the rate of counting for the considered model of an ideal photon detector makes

$$p(t) = \frac{dw}{dt} = \sum_{mn} s_{mn} G_{mn}^{(1,1)}(x,t;x,t) \tag{13}$$

The problem of correlation of modes with different polarizations is a complicated one from the point of view of quantum measurements. So in most cases theoretical consideration goes to the presence of polarization filter. For such case the correlation (13) takes the form

$$p(t) = sG^{(1,1)}(x,t;x,t) \equiv s\langle E^{(-)}(x,t) E^{(+)}(x,t) \rangle , \qquad \hat{E}(x,t) \equiv \hat{E}_n(x,t) e_n \tag{14}$$

confirming that an ideal detector measures a correlation function of the first order with coinciding space-time arguments, i.e. field intensity in a fixed point (e_n is polarization vector depending on the filter).

Correlation properties of radiation manifest themselves in interference experiments. The well-known Young scheme with signals from two apertures interfering can be analyzed in quantum terms. Schematically, we regard (in accordance with Huygens-Fresnel principle) a field value in an observation point x at some time t as a linear combination of field parameters in aperture points x_1 and x_2 at proper time moments. Using our previous considerations concerning quantum detectors, we put down, for example, for negative-frequency part of the electric field strength for a fixed field polarization

$$\hat{E}^{(-)}(x,t) = \alpha_1 \hat{E}^{(-)}(x_1,t_1) + \alpha_2 \hat{E}^{(-)}(x_2,t_2) \tag{15}$$

where $t_{1,2} = t - s_{1,2} / c$ and $s_{1,2} = |x_{1,2} - x|$; α_1 and α_2 are determined by the system geometry. Thus for readings of an ideal detector placed in x we obtain an expression including an interference term

$$2\,\mathrm{Re}\,\alpha_1 \alpha_2^* \langle \hat{E}^{(-)}(x_1,t_1) \hat{E}^{(+)}(x_2,t_2) \rangle .$$

The most important conclusion at this stage is possibility of measuring a correlation function of the first order defined by (8) with arbitrary arguments on the basis of the Young scheme and one photon detector. The stability of the statistical situation is suggested, thus function (8) is transformed into the function of $t_1' - t_1$. So, using polarization filters after apertures, we obtain a scheme for measuring a correlation function (8) in the most general form.

We see that optical measurements with one quantum detector lead to considering a correlation function of the first order (8) with necessity. In order to obtain information about more complex correlation properties of electromagnetic fields, we should consider a more complicated model problem corresponding to the scheme of the famous pioneer experiments of Hanbury Brown and Twiss (Hanbury Brown & Twiss, 1956). We suppose that two ideal detectors of photons are located in points x_1 and x_2; optical shutters are placed in front of the detectors. The shutters are opened at the time moment t_0 and closed at the moments t_1 and t_2. Calculation of probability of photon absorption in each detector gives the following result

$$w^{(2)} = \int_{t_0}^{t_1} d\tau_1 \int_{t_0}^{t_2} d\tau_2 \int_{t_0}^{t_1} d\tau_1' \int_{t_0}^{t_2} d\tau_2' R_{m_1 n_1}(\tau_1 - \tau_1') R_{m_2 n_2}(\tau_2 - \tau_2') G^{(2,2)}_{m_1 m_2, n_1 n_2}(x_1, \tau_1; x_2, \tau_2; x_1, \tau_1'; x_2, \tau_2') \quad (16)$$

where $R_{mn}(\tau - \tau')$ is a sensitivity correlation function determined by (12) and a correlation function of the second order

$$G^{(2,2)}_{m_1 m_2, n_1 n_2}(y_1, y_2; y_1', y_2') = \langle \hat{E}^{(-)}_{m_1}(y_1) \hat{E}^{(-)}_{m_2}(y_2) \hat{E}^{(+)}_{n_1}(y_1') \hat{E}^{(+)}_{n_2}(y_2') \rangle \quad (17)$$

is introduced (we use here an abbreviated notation $y = (x, t)$). In the above-considered case of a broadband detector the rate of coinciding of photon registrations by two detectors makes

$$p^{(2)} = \frac{\partial^2 w^{(2)}}{\partial t_1 \partial t_2} = s_{m_1 n_1} s_{m_2 n_2} G^{(2,2)}_{m_1 m_2, n_1 n_2}(x_1, t_1; x_2, t_2; x_1, t_1; x_2, t_2) \quad (18)$$

with detector parameters s_{mn} introduced in (12). Therefore the Hanbury Brown–Twiss experimental scheme with registering the coincidence of photon absorption by two detectors obtaining signals from the divided light beam with a delay line in front of one of detectors provides measuring of the correlation function of the second order (17) if each detector operates with a certain polarization of the wave.

Generalizations of the Hanbury Brown–Twiss coincidence scheme for the case of *N* detectors are considered as obvious. The rate of *N*-fold coincidences is connected with a correlation function of *N*th order. The analysis of ideal quantum photon detector operation and coincidence scheme by Glauber has elucidated the nature of field functions measured via using the noted schemes – they are functions built with the set of normally ordered operators

$$\langle \hat{E}^{(-)}_{m_1}(y_1) \ldots \hat{E}^{(-)}_{m_M}(y_M) \hat{E}^{(+)}_{n_1}(y_1) \ldots \hat{E}^{(+)}_{n_M}(y_M) \rangle \quad (19)$$

in the case of M detectors. At last, the most general set of normally ordered correlation functions introduced by Glauber (Glauber, 1963) looks like

$$G^{(M,N)}_{m_1...m_M,n_1...n_N}(y_1...y_M,y'_1...y'_N) \equiv \langle \hat{E}^{(-)}_{m_1}(y_1)...\hat{E}^{(-)}_{m_M}(y_M)\hat{E}^{(+)}_{n_1}(y'_1)...\hat{E}^{(+)}_{n_N}(y'_N)\rangle \,. \tag{20}$$

Functions (20) equal to zero usually at $M \neq N$ except very special states with broken symmetry (Glauber, 1969). Such function complex provides the most full description of the field correlation properties. In this picture taking into account magnetic field amplitudes is not necessary since they are simply connected with electric field amplitudes for each mode of electromagnetic field. Notice that the electric-dipole mechanism of absorption really dominates in experiments.

Method of photon counting corresponds to the general ideas of statistical approach; in its terms a number of quantum optics phenomena is described adequately, so the term "quantum optics" is used mainly as "statistical optics". Traditional terminology concerning correlation properties of light is based on the notion "coherence". In scientific literature coherences of the first and second orders are distinguished. It can be substantiated that, for example, the visibility of interference fringes in the Young scheme is determined by the coherence function of the first order that is a normalized correlation function of the first order (Scully & Zubairy, 1997)

$$g^{(1)}(x_1,x_2,\tau) = \frac{\langle \hat{E}^{(-)}(x_1,t)\hat{E}^{(+)}(x_2,t+\tau)\rangle}{\sqrt{\langle \hat{E}^{(-)}(x_1,t)\hat{E}^{(+)}(x_1,t)\rangle\langle \hat{E}^{(-)}(x_2,t+\tau)\hat{E}^{(+)}(x_2,t+\tau)\rangle}} \,. \tag{21}$$

Similarly to (21), the photon grouping effect is determined by the coherence function of the second order

$$g^{(2)}(x,\tau) = \frac{\langle \hat{E}^{(-)}(x,t)\hat{E}^{(-)}(x,t+\tau)\hat{E}^{(+)}(x,t)\hat{E}^{(+)}(x,t+\tau)\rangle}{\langle \hat{E}^{(-)}(x,t)\hat{E}^{(+)}(x,t)\rangle\langle \hat{E}^{(-)}(x,t+\tau)\hat{E}^{(+)}(x,t+\tau)\rangle} \tag{22}$$

Coherences of higher orders (Bogolyubov (Jr.) et al., 1988) can be introduced in the same way. We shall refer to Glauber functions (20) as the main means of field description in quantum optics. Differences between time arguments play the decisive role in the physical interpretation of functions. Taking into account all difficulties and conditions for measurements, functions of lower orders are really urgent for experimental work.

4. Superfluorescence in Dicke model as an important example of collective quantum phenomena

The Dicke model of a system of great quantity of two-level emitters interacting via electromagnetic field (Dicke, 1954) is a noticeable case of synergetics in statistical system behavior during the relaxation processes. Its research history is very informative. R. Dicke came to the conclusion about superradiant state formation proceeding from the analysis of symmetry of quantum states of emitters described with quasispin operators. For long time equilibrium properties of the Dicke model were under discussion and the possibility of phase transition has been established; it was associated with field states in lasers. At the next step it has become clear that self-organizing takes place in the dynamical process and presents some kind of a "dynamical phase transition" (Bogolyubov (Jr.) & Shumovsky, 1987). N excited atoms come to coordinated behavior without the mechanism of stimulated emission and a peak of intensity, proportional to N^2, appeared for modes that were close to the resonant one in a direction determined by the geometry of the system (Banfi & Bonifacio, 1975). So we have a way of coherent generation that is alternative to the laser one. This way can be used hypothetically in X- and γ-ray generators opening wide possibilities for physics and technology.

Collective spontaneous emission in the Dicke quasispin model proved to be one of the most difficult for experimental observations collective quantum phenomena. That is why taking into account real conditions of the experiment is of great importance. Thus great quantity of Dicke model generalizations has been considered. There are two factors dependent of temperature, namely the own motion of emitters and their interaction with the media. The both factors are connected with additional chaotic motion, thus they worsen the prospects of self-organizing in a system. The last factor is discussed traditionally as an influence of a cavity (resonator) since experiments in superradiance use laser technology (Kadantseva et al., 1989). The corresponding theoretical analysis is based on modeling the cavity with a system of oscillators (Louisell, 1964). The problem of influence of emitter motion (which is of different nature in different media) can be solved with taking into account this motion via a nonuniform broadening of the working frequency of emitters (Bogolyubov (Jr.) & Shumovsky, 1987). The dispersion of emitter frequencies results in an additional fading in a system and elimination of singularities in kinetic coefficients.

Traditional investigations obtain conclusions about a superfluorescent impulse generation on the basis of calculated behavior of the system of two-level emitters. The problem of light generation in the Dicke model can be investigated in the framework of the Bogolyubov method of eliminating boson variables (Bogolyubov (Jr.) & Shumovsky, 1987) with the suggestion of equilibrium state of field with a certain temperature. The correlation properties of light remain unknown in such picture. Good results can be obtained by applying the Bogolyubov reduced description method (Lyagushyn et al., 2005) to the model. The reduced description method eliminates some difficulties in the Dicke model investigations and allows both to take into account some additional factors (the orientation and motion of emitters, for instance) and to introduce more detailed description of the field. A kind of correlation functions to be used in such approach will be of interest for us.

5. Quantum models for electromagnetic field in media

The main problem of quantum optics is diagnostics of electromagnetic field (f -system) interacting with a medium (m -system). In this connection we have considered a number of models of medium and medium-field interaction. From various points of view the Dicke model of medium consisting of two-level emitters is very useful for such analysis. In the Coulomb gauge it is described by the Hamilton operator (Lyagushyn & Sokolovsky, 2010b)

$$\hat{H} = \hat{H}_{\mathrm{f}} + \hat{H}_{\mathrm{m}} + \hat{H}_{\mathrm{mf}}, \quad \hat{H}_{\mathrm{f}} \equiv \sum_{k\alpha} \hbar\omega_k c^{+}_{k\alpha} c_{k\alpha}, \quad \hat{H}_{\mathrm{m}} = \hbar\omega \sum_{1\le a\le N} \hat{r}_{az}, \tag{23}$$

$$\hat{H}_{\mathrm{mf}} = -\int d^3x \hat{E}^t_n(x) \hat{P}_n(x)$$

Here $\hat{r}_{an}$ is a quasispin operator, a is emitter's number, α is polarization index, $\hat{P}_n(x)$ is the density of electric dipole moment (polarization) of emitters

$$\hat{P}_n(x) = 2 \sum_{1 \le a \le N} d_{an} \hat{r}_{ax} \delta(x - x_a) . \tag{24}$$

We neglect emitter-emitter interaction in (23). Operators of vector potential, transversal electric field and magnetic field are expressed via creation and annihilation boson operators $c^{+}_{k\alpha}, c_{k\alpha}$ by formulas (9), (10) and commutation relations

$$[\hat{E}^t_m(x), \hat{E}^t_n(x')] = 0 , \quad [\hat{B}_m(x), \hat{B}_n(x')] = 0 , \quad [\hat{B}_m(x), \hat{E}^t_n(x')] = \varepsilon_{mnl} 4\pi i \hbar c \frac{\partial \delta(x - x')}{\partial x_l} \tag{25}$$

are valid (we use the notation $\hat{E}^t_n(x)$ for electric field operator (10) in the discussion of the field-emitters system).

It is very convenient to use operator evolution equations for investigating the dynamics of the system (23). The Maxwell operator equations have a known form

$$\dot{\hat{E}}_n(x) = c \operatorname{rot}_n \hat{B}(x) - 4\pi \hat{j}_n(x), \qquad \dot{\hat{B}}_n(x) = -c \operatorname{rot}_n \hat{E}(x) \tag{26}$$

where total electric field and electromagnetic current

$$\hat{E}_n(x) \equiv \hat{E}^t_n(x) - 4\pi \hat{P}_n(x), \quad \hat{j}_n(x) \equiv \dot{\hat{P}}_n(x) = -2\omega \sum_a d_{an} \hat{r}_{ay} \delta(x - x_a) \tag{27}$$

are introduced. Energy density of emitter medium

$$\hat{\varepsilon}(x) = \hbar\omega \sum_{1 \le a \le N} \hat{r}_{az} \delta(x - x_a)$$

obeys the evolution equation

$$\dot{\hat{\varepsilon}}(x) = \hat{j}_n(x) \hat{E}^t_n(x) \tag{28}$$

which describes the Joule heat exchange between the emitters and field. Since the field parameters are considered in different spatial points, we obtain the possibility of investigating the field correlation properties.

Also the model of electromagnetic field in plasma medium plays a significant role. The Hamilton operator of such system in the Coulomb gauge was taken in the paper (Sokolovsky & Stupka, 2004) in the form

$$\hat{H} = \hat{H}_{\mathrm{f}} + \hat{H}_{\mathrm{m}} + \hat{H}_{\mathrm{mf}} , \qquad \hat{H}_{\mathrm{f}} \equiv \sum_{k\alpha} \hbar \omega_k c^{+}_{k\alpha} c_{k\alpha} , \qquad \hat{H}_{\mathrm{mf}} = \hat{H}_1 + \hat{H}_2 , \tag{29}$$

$$\hat{H}_1 = -\frac{1}{c} \int dx \hat{A}_n(x) \hat{j}_n(x) , \qquad \hat{H}_2 = \frac{1}{2c^2} \int dx \hat{A}^2(x) \hat{\chi}(x) \qquad \left(\hat{\chi}(x) \equiv \sum_a \frac{e_a^2}{m_a} \hat{n}_a(x) \right).$$

Here $\hat{H}_m$ is the Hamilton operator of plasma particles with account of Coulomb interaction, $\hat{j}_n(x)$ is electric current, $\hat{n}_a(x)$ is density operator of the a th component of the system.

6. Reduced description of electromagnetic field in medium. Role of field correlations

Here we discuss kinetics of electromagnetic field in a medium. This theory must connect dynamics of the field with dynamics of the medium. The problem can be solved only on the basis of the reduced description of a system. One has to choose a set of microscopic quantities in such way that their average values describe the system completely. Therefore, the Bogolyubov reduced description method (Akhiezer & Peletminskii, 1981) can be a basis for the general consideration of the problem. In this approach its starting point is a quantum Liouville equation for the statistical operator $\rho(t)$ of a system including electromagnetic field and a medium

$$\partial_t \rho(t) = -\frac{i}{\hbar}[\hat{H}, \rho(t)], \qquad \hat{H} = \hat{H}_{\mathrm{f}} + \hat{H}_{\mathrm{m}} + \hat{H}_{\mathrm{mf}}. \tag{30}$$

The method is based on the functional hypothesis describing a structure of the operator $\rho(t)$ at large times (Bogolyubov, 1946)

$$\rho(t) \xrightarrow[t \gg \tau_0]{} \rho(\xi(t,\rho_0), \eta(t,\rho_0)) \equiv \rho^{(+)}(t) \qquad (\rho_0 \equiv \rho(t=0)) \tag{31}$$

where reduced description parameters of the field $\xi_\mu(t,\rho_0)$ and matter $\eta_a(t,\rho_0)$ are defined in a natural way

$$\xi_\mu(t,\rho_0) = \mathrm{Sp}\,\rho^{(+)}(t)\hat{\xi}_\mu, \qquad \eta_a(t,\rho_0) = \mathrm{Sp}\,\rho^{(+)}(t)\hat{\eta}_a \tag{32}$$

(τ_0 is a characteristic time determined by an initial state of the system ρ_0 and a used set of reduced description parameters). The set of parameters $\xi_\mu(t,\rho_0)$, $\eta_a(t,\rho_0)$ is determined by the possibilities and traditions of experiments as well as by theoretical considerations (for simplicity we will drop ρ_0 in the parameters). The development of the problem investigation has resulted in finding the main approximation for the statistical operator $\rho(\xi,\eta)$, so called a quasiequilibrium statistical operator $\rho_q(Z(\xi), X(\eta))$ (though it describes states which are far from the equilibrium) defined by the relations

$$\rho_q(Z, Z_m) = \rho_{\mathrm{f}}(Z)\rho_{\mathrm{m}}(Z_{\mathrm{m}}); \tag{33}$$

$$\rho_{\mathrm{f}}(Z) = \exp\{\Phi(Z) - \sum_\mu Z_\mu \hat{\xi}_\mu\}, \quad \mathrm{Sp}_{\mathrm{f}}\,\rho_{\mathrm{f}}(Z) = 1, \quad \mathrm{Sp}_{\mathrm{f}}\,\rho_{\mathrm{f}}(Z(\xi))\hat{\xi}_\mu = \xi_\mu; \tag{34}$$

$$\rho_{\mathrm{m}}(X) = \exp\{\Omega(X) - \sum_a X_a \hat{\eta}_a\}, \quad \mathrm{Sp}_{\mathrm{m}}\,\rho_{\mathrm{m}}(X) = 1, \quad \mathrm{Sp}_{\mathrm{m}}\,\rho_{\mathrm{m}}(X(\eta))\hat{\eta}_a = \eta_a. \tag{35}$$

According to the common idea, electromagnetic field in medium is usually described by average values of electric $E_n(x,t)$ and magnetic $B_n(x,t)$ fields. So, it seems possible to choose operators $\hat{\xi}_\mu$ in (32) as $\hat{\xi}_{\alpha n}(x)$: $\hat{\xi}_{1n}(x)=\hat{E}^t_n(x)$, $\hat{\xi}_{2n}(x)=\hat{B}_n(x)$. However, in this case the statistical operator $\rho_{\mathrm{f}}(Z)$ does not exist (its exponent contains only linear in Bose amplitudes form and $\rho_{\mathrm{f}}(Z)$ is non-normalized). Therefore, one has to use a wider set of parameters $\hat{\xi}_\mu$ in conformity with the observation made in (Peletminkii et al., 1975). At least, exponent in (34) should contain quadratic terms. So the simplest quasiequilibrium statistical operator of the field can be written as

$$\rho_{\mathrm{f}}(Z)=\exp\{\Phi(Z)-\sum_{\alpha k,\alpha' k'} Z^{\alpha\alpha'}_{kk'}c^+_{\alpha k}c_{\alpha' k'}-(\sum_{\alpha k,\alpha' k'}\tilde{Z}^{\alpha\alpha'}_{kk'}c^+_{\alpha k}c^+_{\alpha' k'}+\sum_{\alpha k}Z^\alpha_k c^+_{\alpha k}+h.c.)\} \tag{36}$$

Kinetics of the field based on this statistical operator describes states with zero average fields at $Z^\alpha_k=0$. Quadratic terms in (36) correspond to binary fluctuation of the field $\langle\xi^x_{\alpha m}\xi^{x'}_{\alpha' n}\rangle_t$ (or two binary correlations $(\xi^x_{\alpha m}\xi^{x'}_{\alpha' n})_t$) as additional reduced description parameters

$$\mathrm{Sp}\,\rho^{(+)}(t)\frac{1}{2}\{\hat{\xi}_{\alpha m}(x),\hat{\xi}_{\alpha' n}(x')\}=\langle\xi^x_{\alpha m}\xi^{x'}_{\alpha' n}\rangle_t\,,\qquad (\xi^x_{\alpha m}\xi^{x'}_{\alpha' n})_t=\langle\xi^x_{\alpha m}\xi^{x'}_{\alpha' n}\rangle_t-\xi_{\alpha m}(x,t)\xi_{\alpha' n}(x',t) \tag{37}$$

In other words, the quasiequilibrium statistical operator (34) corresponds to field description by average values of operators

$$\hat{\xi}_\mu:\quad \hat{\xi}_{\alpha n}(x)\,,\quad \frac{1}{2}\{\hat{\xi}_{\alpha m}(x),\hat{\xi}_{\alpha' n}(x')\}\,. \tag{38}$$

The theory can be significantly simplified in the Peletminskii-Yatsenko model (Akhiezer & Peletminskii, 1981) in which

$$\frac{1}{\hbar}[\hat{H}_{\mathrm{f}},\hat{\xi}_\mu]=\sum_{\mu'}c_{\mu\mu'}\hat{\xi}_{\mu'}\,,\qquad \frac{1}{\hbar}[\hat{H}_{\mathrm{m}},\hat{\eta}_a]=\sum_{aa'}c_{aa'}\hat{\eta}_{a'} \tag{39}$$

where $c_{\mu\mu'}$, $c_{aa'}$ are some coefficients. Operators of electromagnetic field $\hat{E}^t_n(x)$, $\hat{B}_n(x)$ and operator $\hat{\varepsilon}(x)$ satisfy these conditions

$$[\hat{H}_{\mathrm{f}},\hat{E}^t_n(x)]=-ic\hbar\,\mathrm{rot}_n\,\hat{B}(x)\,,\qquad [\hat{H}_{\mathrm{f}},\hat{B}_n(x)]=ic\hbar\,\mathrm{rot}_n\,\hat{E}^t(x)\,, \tag{40}$$

$$[\hat{H}_{\mathrm{m}},\hat{\varepsilon}(x)]=0\,,$$

therefore, relations (39) are valid for all field operators in (38).

In usual kinetic theory nonequilibrium states of quantum system are described by one-particle density matrix $n^{\alpha\alpha'}_{kk'}(t)$

$$n^{\alpha\alpha'}_{kk'}(t)\equiv\mathrm{Sp}\,\rho^{(+)}(t)c^+_{\alpha k}c_{\alpha' k'}\,. \tag{41}$$

States, for which parameters

$$\tilde{n}^{\alpha\alpha'}_{kk'}(t)\equiv\mathrm{Sp}\,\rho^{(+)}(t)c_{\alpha k}c_{\alpha' k'}\,,\qquad x_{\alpha k}(t)\equiv\mathrm{Sp}\,\rho^{(+)}(t)c_{\alpha k} \tag{42}$$

are not equal to zero, are considered as states with a broken symmetry. Therefore, $\tilde{n}_{kk'}^{\alpha\alpha'}(t)$ is called an anomalous one-particle density matrix. However, average electromagnetic fields are expressed through $x_{\alpha k}(t)$. Instead of density matrices Wigner distribution functions are widely used (de Groot, S. & Suttorp L., 1972)

$$\mathrm{f}_k^{\alpha\alpha'}(x,t) \equiv \mathrm{Sp}\,\rho^{(+)}(t)\hat{\mathrm{f}}_k^{\alpha\alpha'}(x), \qquad \tilde{\mathrm{f}}_k^{\alpha\alpha'}(x,t) \equiv \mathrm{Sp}\,\rho^{(+)}(t)\hat{\tilde{\mathrm{f}}}_k^{\alpha\alpha'}(x) \tag{43}$$

where

$$\hat{\mathrm{f}}_k^{\alpha\alpha'}(x) = \sum_q c_{\alpha,k-q/2}^{+} c_{\alpha',k+q/2} e^{iqx}, \qquad \hat{\tilde{\mathrm{f}}}_k^{\alpha\alpha'}(x) = \sum_q c_{\alpha,k+q/2} c_{\alpha',-k+q/2} e^{iqx}. \tag{44}$$

Simple relations between average field, correlations of the field, density matrices and Wigner distribution functions can be established by the formula

$$c_{\alpha k} = (8\pi\omega_k \hbar V)^{-1/2} e_{\alpha kn}^{*} \int d^3x \{\hat{Z}_n(x)/k - i\hat{E}_n^t(x)\} e^{-ikx}, \qquad \hat{Z}_n(x) \equiv \mathrm{rot}_n \hat{B}(x). \tag{45}$$

Further on kinetics of electromagnetic field in medium consisting of two-level emitters with the Hamilton operator (23) is considered in more detail. According to the general theory (Akhiezer & Peletminskii, 1981), an integral equation for the statistical operator $\rho(\xi,\varepsilon)$ introduced by the functional hypothesis (31) can be obtained (Lyagushyn & Sokolovsky, 2010b)

$$\rho(\xi,\varepsilon) = \rho_{\mathrm{f}}(Z(\xi))\rho_{\mathrm{m}}(X(\varepsilon)) + \int_{-\infty}^{0} d\tau e^{\frac{i}{\hbar}\tau\hat{H}_0} \Big\{ \frac{i}{\hbar}[\rho(\xi,\varepsilon),\hat{H}_{\mathrm{mf}}] - \tag{46}$$

$$- \sum_{\mu} \frac{\partial \rho(\xi,\varepsilon)}{\partial \xi_\mu} M_\mu(\xi,\varepsilon) - \int d^3x \frac{\delta\rho(\xi,\varepsilon)}{\delta\varepsilon(x)} M(x,\xi,\varepsilon) \Big\}_{\xi \to e^{-ic\tau}\xi} e^{-\frac{i}{\hbar}\tau\hat{H}_0}$$

where functions $M_\mu(\xi,\varepsilon)$, $M(\xi,\varepsilon)$ are defined as right-hand sides of evolution equations for the reduced description parameters

$$\partial_t \xi_\mu(t) = i\sum_{\mu'} c_{\mu\mu'}\xi_{\mu'}(t) + M_\mu(\xi(t),\varepsilon(t)), \qquad \partial_t \varepsilon(x,t) = M(x,\xi(t),\varepsilon(t)); \tag{47}$$

$$M_\mu(\xi,\varepsilon) \equiv \frac{i}{\hbar}\mathrm{Sp}\,\rho(\xi,\varepsilon)[\hat{H}_{\mathrm{mf}},\hat{\xi}_\mu], \qquad M(x,\xi,\varepsilon) \equiv \frac{i}{\hbar}\mathrm{Sp}\,\rho(\xi,\varepsilon)[\hat{H}_{\mathrm{mf}},\hat{\varepsilon}(x)]$$

(see notations in (39)). Quasiequilibrium statistical operator of the emitters

$$\rho_{\mathrm{m}}(X) = w_d(d) w_\sigma(\omega) \exp\{\Omega(X) - \int d^3x X(x)\hat{\varepsilon}(x)\} \tag{48}$$

describes a state of local equilibrium of the emitter medium with temperature $T(x) = X(x)^{-1}$ in the considered case. Function $w_d(d)$ describes distribution of orientations of emitter dipole moments (Lyagushyn et al., 2008). Further it is assumed for simplicity that correlations of dipole orientations are absent and their distribution is isotropic one. Function $w_\sigma(\omega)$ is defined by formulas

$$w_\sigma(\omega) = c(\sigma)\frac{\sigma}{(\omega-\omega_0)^2+\sigma^2}\,, \qquad \int_0^{+\infty} d\omega w_\sigma(\omega) = 1 \qquad (\sigma \ll 1) \tag{49}$$

and phenomenologically accounts for non-resonant interaction between the field and emitters.

The obtained integral equation is solved in perturbation theory in emitter-field interaction $\hat{H}_{\mathrm{mf}} \sim \lambda$ ($\lambda \ll 1$). Important convenience is provided by the structure of $\rho_{\mathrm{f}}(Z(\xi))$ allowing to use the Wick--Bloch--de Dominicis theorem. However, one needs this theorem for calculating contributions of the third and higher orders of the perturbation theory to the statistical operator $\rho(\xi,\varepsilon)$. Averages that are linear and bilinear in the field can be calculated on the basis of relations:

$$\mathrm{Sp}_{\mathrm{f}}\rho_{\mathrm{f}}(Z(\xi))\hat{\xi}_{\alpha n}(x) = \xi_{\alpha n}(x)\,, \tag{50}$$

$$\mathrm{Sp}_{\mathrm{f}}\rho_{\mathrm{f}}(Z(\xi))\hat{\xi}_{\alpha m}(x)\hat{\xi}_{\alpha' n}(x') = (\xi^{x}_{\alpha m}\xi^{x'}_{\alpha' n}) + \xi_{\alpha m}(x)\xi_{\alpha' n}(x') + \frac{1}{2}[\hat{\xi}_{\alpha m}(x), \hat{\xi}_{\alpha' n}(x')]\,.$$

Moreover, according to the general theory of the Peletminskii-Yatsenko model (Akhiezer & Peletminskii, 1981) the same formulas are valid for calculations with the statistical operator $\rho(\xi,\varepsilon)$:

$$\mathrm{Sp}\rho(\xi,\varepsilon)\hat{\xi}_{\alpha n}(x) = \xi_{\alpha n}(x)\,, \tag{51}$$

$$\mathrm{Sp}\rho(\xi,\varepsilon)\hat{\xi}_{\alpha m}(x)\hat{\xi}_{\alpha' n}(x') = (\xi^{x}_{\alpha m}\xi^{x'}_{\alpha' n}) + \xi_{\alpha m}(x)\xi_{\alpha' n}(x') + \frac{1}{2}[\hat{\xi}_{\alpha m}(x), \hat{\xi}_{\alpha' n}(x')]\,.$$

Averages with a quasiequilibrium statistical operator of the medium are calculated by the method developed for spin systems (Lyagushyn et al., 2005). It gives, for example, an expression for energy density of emitter medium via its temperature $T(x)$ and density $n(x)$

$$\varepsilon(x) = -\frac{\hbar\omega}{2} n(x) \mathrm{th}\frac{\hbar\omega}{2T(x)} \qquad (\, n(x) \equiv \sum_{1\le a\le N} \delta(x - x_a)\,). \tag{52}$$

Integral equation (46) solution gives evolution equations for all parameters of the reduced description. Average electric and magnetic fields satisfy the Maxwell equations

$$\partial_t E_n(x,t) = c\,\mathrm{rot}_n\, B(x,t) - 4\pi J_n(x,\xi(t),\varepsilon(t))\,, \qquad \partial_t B_n(x,t) = -c\,\mathrm{rot}_n\, E(x,t) \tag{53}$$

where average current density in terms of the total electric field is given by the relation

$$J_n(x,\xi,\varepsilon) = \int dx' \sigma(x - x', \varepsilon(x)) E_n(x') + c\int dx' \chi(x - x', \varepsilon(x)) Z_n(x') + O(\lambda^3) \tag{54}$$

$$E_n(x,t) = E^t_n(x,t) + O(\lambda^2)\,, \qquad Z_n(x,t) \equiv \mathrm{rot}_n B(x,t)$$

(for all parameters $A(\xi,\varepsilon)\equiv \mathrm{Sp}\,\rho(\xi,\varepsilon)\hat{A}$). This material equation takes into account spatial dispersion and Fourier transformed functions $\sigma(x,\varepsilon)$, $\chi(x,\varepsilon)$ give conductivity $\sigma(k,\varepsilon)$ and magnetic susceptibility $\chi(k,\varepsilon)$ of the emitter medium

$$\sigma(k,\varepsilon)=-\frac{2\pi}{3}\frac{\varepsilon d^2}{\hbar^2}w_\sigma(\omega_k), \quad \chi(k,\varepsilon)=-\frac{4}{3}\frac{\varepsilon d^2}{\hbar^2}\int_0^{+\infty}d\omega\, w_\sigma(\omega)\mathrm{P}\frac{1}{\omega^2-\omega_k^2}. \tag{55}$$

Average density of the dipole moment of emitters is given by expression

$$P_n(x,\varepsilon,\gamma)=\int dx'\kappa(x-x',\varepsilon(x))E_n(x')++c\int dx'\alpha(x-x',\varepsilon(x))Z_n(x')+O(\lambda^3) \tag{56}$$

where

$$\kappa(k,\varepsilon)=\chi(k,\varepsilon), \qquad \alpha(k,\varepsilon)=-\sigma(k,\varepsilon)/\omega_k^2. \tag{57}$$

Evolution equation for energy density $\varepsilon(x,t)$ of emitters has the form

$$\partial_t\varepsilon(x,t)=L(x,\xi(t),\varepsilon(t)), \tag{58}$$

$$L(x,\xi,\varepsilon)=\int dx'\sigma(x-x',\varepsilon(x))\{(E_n^xE_n^{x'})+E_n(x)E_n(x')\}+$$

$$+c\int dx'\chi(x-x',\varepsilon(x))\{(E_n^xB_n^{x'})+E_n(x)B_n(x')\}+R(n(x))+O(\lambda^3).$$

The last term describes dipole radiation of the emitters

$$R(n)\equiv-\frac{2d^2}{3\pi c^3}n\int_0^{+\infty}d\omega\,\omega^4 w_\sigma(\omega) \tag{59}$$

and for small σ gives a known expression

$$R(n)=-\frac{2d^2\omega_0^4}{3\pi c^3}n\,. \tag{60}$$

Evolution equations for correlation functions of electromagnetic field in terms of the total electric field can be written in the form

$$\partial_t(E_m^xE_n^{x'})_t=c\,\mathrm{rot}_m(B^xE_n^{x'})_t+c\,\mathrm{rot}'_n(E_m^xB^{x'})_t-4\pi(J_m^xE_n^{x'})_t-4\pi(E_m^xJ_n^{x'})_t, \tag{61}$$

$$\partial_t(E_m^xB_n^{x'})_t=c\,\mathrm{rot}_m(B^xB_n^{x'})_t-c\,\mathrm{rot}'_n(E_m^xE^{x'})_t-4\pi(J_m^xB_n^{x'})_t,$$

$$\partial_t(B_m^xE_n^{x'})_t=-c\,\mathrm{rot}_m(E^xE_n^{x'})_t+c\,\mathrm{rot}'_n(B_m^xB^{x'})_t-4\pi(B_m^xJ_n^{x'})_t,$$

$$\partial_t(B_m^xB_n^{x'})_t=-c\,\mathrm{rot}_m(E^xB_n^{x'})_t-c\,\mathrm{rot}'_n(B_m^xE^{x'})_t.$$

Current-field correlation functions are defined analogously to (37). Material equations for these correlations are given by expressions in terms of the total electric field

$$(E_m^x J_n^{x'})_t = \int dx'' \sigma(x'-x'', \varepsilon(x',t))(E_m^x E_n^{x''})_t + \tag{62}$$

$$+c\int dx'' \chi(x'-x'', \varepsilon(x',t))(E_m^x Z_n^{x''})_t + S_{mn}(x-x', n(x')) + O(\lambda^3),$$

$$(B_m^x J_n^{x'})_t = \int dx'' \sigma(x'-x'', \varepsilon(x',t))(B_m^x E_n^{x''})_t +$$

$$+c\int dx'' \chi(x'-x'', \varepsilon(x',t))(B_m^x Z_n^{x''})_t + T_{mn}(x-x', n(x')) + O(\lambda^3),$$

where Fourier transformed functions $S_{mn}(x,n)$, $T_{mn}(x,n)$ are given by expressions

$$S_{mn}(k,n) = -\frac{2\pi}{3} d^2 n(\delta_{mn} - \tilde{k}_m \tilde{k}_n)\omega_k^2 w_\sigma(\omega_k), \tag{63}$$

$$T_{mn}(k,n) = \frac{4\pi i}{3} c d^2 n \, e_{mnl} k_l \int_0^{+\infty} d\omega \, w_\sigma(\omega) \mathrm{P}\frac{\omega}{\omega^2 - \omega_k^2}$$

Quantities $S_{mn}(k,n)$, $T_{mn}(k,n)$ determine equilibrium correlations of the electromagnetic field. Comparing relations (54) and (62) shows that the Onsager principle is valid for the considered system.

Hereafter we consider kinetics of electromagnetic field in plasma medium with the Hamiltonian (29) in more detail. We restrict ourselves by considering equilibrium plasma (Sokolovsky & Stupka, 2004) and states of the field described by average fields $E_n^t(x,t)$, $B_n(x,t)$ and one-particle density matrix $n_{kk'}^{\alpha\alpha'}(t)$ defined in (41). The problem for plasma medium in terms of hydrodynamic states has been investigated in (Sokolovsky & Stupka, 2005). Instead of average fields and matrix $n_{kk'}^{\alpha\alpha'}(t)$ one can use average Bose amplitudes $x_{\alpha k}(t)$ defined in (42) and correlation function

$$g_{kk'}^{\alpha\alpha'}(t) = n_{kk'}^{\alpha\alpha'}(t) - x_{\alpha k}^*(t) x_{\alpha' k'}(t) \,. \tag{64}$$

So, for this problem in above notations we have parameters ξ_μ: $n_{kk'}^{\alpha\alpha'}$, $x_{\alpha k}$, $x_{\alpha k}^*$ and corresponding operators $\hat{\xi}_\mu$: $c_{\alpha k}^+ c_{\alpha' k'}$, $c_{\alpha k}$, $c_{\alpha k}^+$. A statistical operator of the system introduced by the functional hypothesis depends in this case only on the field variables and satisfies the integral equation

$$\rho(\xi) = \rho_{\mathrm{f}}(Z(\xi)) w_{\mathrm{m}} + \int_{-\infty}^{0} d\tau e^{\frac{i}{\hbar}\tau \hat{H}_0} \left\{ \frac{i}{\hbar}[\rho(\xi), \hat{H}_{\mathrm{mf}}] - \sum_\mu \frac{\partial \rho(\xi)}{\partial \xi_\mu} M_\mu(\xi) \right\}_{\xi \to e^{-ic\tau}\xi} e^{-\frac{i}{\hbar}\tau \hat{H}_0}, \tag{65}$$

where quasiequilibrium statistical operator $\rho_{\mathrm{f}}(Z)$ is given by formula (36) with $\tilde{Z}_{kk'}^{\alpha\alpha'} = 0$, w_{m} is a statistical operator of equilibrium plasma

$$w_{\mathrm{m}} = e^{(\Omega-\hat{H}_{\mathrm{m}}+\sum_a \mu_a \hat{N}_a)/T} \qquad (\hat{N}_a \equiv \int dx \hat{n}_a(x)). \tag{66}$$

Functions $M_\mu(\xi)$ define the right-hand sides of evolution equations for the reduced description parameters

$$\partial_t \xi_\mu(t) = i\sum_{\mu'} c_{\mu\mu'}\xi_{\mu'}(t) + M_\mu(\xi(t)), \qquad M_\mu(\xi) \equiv \frac{i}{\hbar}\mathrm{Sp}\,\rho(\xi)[\hat{H}_{\mathrm{mf}}, \hat{\xi}_\mu]. \tag{67}$$

Integral equation (65) is solvable in a perturbation theory in plasma-field interaction based on estimations $\hat{H}_1 \sim \lambda^1$, $\hat{H}_2 \sim \lambda^2$ (see (29)). As a result, evolution equations for the reduced description parameters take the form (Sokolovsky & Stupka, 2004)

$$\partial_t g_{kk'}^{\alpha\alpha'} = i(\Omega_k - \Omega_{k'})g_{kk'}^{\alpha\alpha'} - (\nu_k + \nu_{k'})(g_{kk'}^{\alpha\alpha'} - n_k \delta_{\alpha\alpha'}\delta_{kk'}) + O(\lambda^3), \tag{68}$$

$$\partial_t x_{\alpha k} = -(i\Omega_k + \nu_k)x_{\alpha k} + (\nu_k + i\omega_k \chi_k)x^*_{\alpha,-k} + O(\lambda^3)$$

where Ω_k is photon spectrum in the plasma, n_k is the Planck distribution with the plasma temperature, ν_k is a frequency of photon emission and absorption. These quantities are given by formulas

$$\Omega_k \equiv \omega_k\{1 - 2\pi\chi(k)\}, \quad \nu_k \equiv 2\pi\sigma(k). \tag{69}$$

The second equation in (68) is a form of the Maxwell equations (53) with similar to (54) material equation

$$J_n(x,\xi) = \int dx' \sigma(x-x')E_n(x') + c\int dx' \chi(x-x')Z_n(x') + O(\lambda^3). \tag{70}$$

This material equation takes into account spatial dispersion and Fourier transformed functions $\sigma(x)$, $\chi(x)$ give conductivity $\sigma(k)$ and magnetic susceptibility $\chi(k)$ of the plasma medium. Their values are given by relations

$$\sigma(k) = -\frac{\mathrm{Im}\,G(k,\omega_k)}{\omega_k}, \qquad \chi(k) = -\frac{\mathrm{Re}\,G(k,\omega_k) + \chi}{c\,\omega_k}, \tag{71}$$

where $G(k,\omega)$ is a transversal part of current-current Green function:

$$G(k,\omega) \equiv \frac{1}{2}G_{mn}(k,\omega)(\delta_{mn} - \tilde{k}_m\tilde{k}_n), \qquad G_{mn}(x,t) = -\frac{i}{\hbar}\theta(t)\mathrm{Sp}_{\mathrm{m}}\, w_{\mathrm{m}}[\hat{j}_m(x,t), \hat{j}_n(0)]; \tag{72}$$

$$\chi \equiv \sum_a \frac{n_a e_a^2}{m_a} \equiv \frac{\Omega^2}{4\pi}.$$

In fact, the obtained results are valid for $kc \gg \Omega$ where Ω is Langmuir frequency.

7. Connection between correlation functions of different nature and some suitable representations for them

One can notice that simultaneous correlation functions of field amplitudes of (37) type arise in a natural way in the framework of the reduced description method. At the same time Glauber correlation functions of (19) type (including positive-frequency and negative-frequency parts of the electric field operator (11) in the interaction picture) seem to be observable quantities from the point of view of experimental possibilities. The most interesting effects of quantum optics can be described with non-simultaneous Glauber functions (Lyagushyn & Sokolovsky, 2010a; Lyagushyn et al., 2011). Nevertheless we can insist that there are no real contradictions between the approaches. Correlation functions (19) characterize properties of electromagnetic field described by the statistical operator ρ. In the previous section we have been constructed a reduced description for electromagnetic field in emitter medium and in plasma medium. These theories lead not only to equations for the reduced description parameters but also to the expression for corresponding nonequilibrium statistical operators. For the field-emitters system a nonequilibrium statistical operator has the form

$$\rho(\xi,\varepsilon)=\rho_{\mathrm{f}}(Z(\xi))\rho_{\mathrm{m}}(X(\varepsilon))-\frac{i}{\hbar}\int_{-\infty}^{0}d\tau\int dx\,[\rho_{\mathrm{f}}(Z(\xi))\rho_{\mathrm{m}}(X(\varepsilon)),\hat{E}_n^t(x,\tau)\hat{P}_n(x,\tau)]+O(\lambda^2)\,. \quad (73)$$

where $\hat{E}_n^t(x,\tau)$, $\hat{P}_l(x,\tau)$ are operators $\hat{E}_n^t(x)$, $\hat{P}_n(x)$ in the interaction picture. Analogously, a nonequilibrium statistical operator for the field-plasma system is given by the formula

$$\rho(\xi)=\rho_{\mathrm{f}}(Z(\xi))w_{\mathrm{m}}-\frac{i}{c\hbar}\int_{-\infty}^{0}d\tau\int dx[\rho_{\mathrm{f}}(Z(\xi))w_{\mathrm{m}},\hat{A}_n(x,\tau)\hat{j}_n(x,\tau)]+O(\lambda^2) \quad (74)$$

where $\hat{A}_n(x,\tau)$, $\hat{j}_n(x,\tau)$ are operators $\hat{A}_n(x)$, $\hat{j}_n(x)$ in the interaction picture. According to general theory of the Peletminskii-Yatsenko model (Akhiezer & Peletminskii, 1981), the following relations for the field-emitters system

$$\mathrm{Sp}\,\rho(\xi,\varepsilon)c_{\alpha k}=\mathrm{Sp}\,\rho_{\mathrm{f}}(Z(\xi))c_{\alpha k}\,,\qquad \mathrm{Sp}\,\rho(\xi,\varepsilon)c^{+}_{\alpha k}c_{\alpha' k'}=\mathrm{Sp}\,\rho_{\mathrm{f}}(Z(\xi))c^{+}_{\alpha k}c_{\alpha' k'}\,, \quad (75)$$

$$\mathrm{Sp}\,\rho(\xi,\varepsilon)c_{\alpha k}c_{\alpha' k'}=\mathrm{Sp}\,\rho_{\mathrm{f}}(Z(\xi))c_{\alpha k}c_{\alpha' k'}$$

and for the field-plasma system

$$\mathrm{Sp}\,\rho(\xi)c_{\alpha k}=\mathrm{Sp}_{\mathrm{f}}\,\rho_{\mathrm{f}}(Z(\xi))c_{\alpha k}\,,\qquad \mathrm{Sp}\,\rho(\xi)c^{+}_{\alpha k}c_{\alpha' k'}=\mathrm{Sp}_{\mathrm{f}}\,\rho_{\mathrm{f}}(Z(\xi))c^{+}_{\alpha k}c_{\alpha' k'} \quad (76)$$

are valid. Average of products of three and more Bose operators should be calculated with taking into account the second term in expressions (73), (74) and using the Wick–Bloch–de Dominicis theorem. It is convenient to perform the calculation of correlation functions (23) for the field-plasma system through using formulas (11), (74). For the field-emitters system the following formula

$$\hat{E}_n^{(+)}(x,t)=i\int dx'\{D_{+}(x-x',t)\hat{Z}_n(x')+\frac{1}{c}\dot{D}_{+}(x-x',t)\hat{E}_n^t(x')\} \quad (77)$$

can be useful. Here $D_+(x,t)$ is a standard function widely used in electromagnetic theory (Akhiezer A. & Berestetsky V., 1969) and defined by expression

$$D_+(x,t) \equiv \frac{1}{2}\int \frac{d^3k}{(2\pi)^3}\frac{1}{k}e^{i(kx-\omega_k t)} . \tag{78}$$

Calculation of the simplest correlation function $G_{nl}^{(1,1)}(y_1,y_1')$ can be done according to (75), (76) exactly. For example, for the field-plasma system one has

$$G_{mn}^{(1,1)}(x,t;x',t') = \sum_{k\alpha,k'\alpha'} \frac{2\pi\hbar c}{V}(kk')^{1/2} e^{*}_{\alpha km} e_{\alpha' k' n} e^{i(\omega_k t - kx)} e^{-i(\omega_{k'} t' - k'x')} n_{kk'}^{\alpha\alpha'} \tag{79}$$

An exact expression for this correlation function of the field-emitters system is given by the formula

$$G_{mn}^{(1,1)}(x,t;x',t') = \int dx_1 dx_1' \Big\{ D_+^*(x-x_1,t)D_+(x'-x_1',t')\langle Z_m^{x_1} Z_n^{x_1'}\rangle + \tag{80}$$

$$+\frac{1}{c^2}\dot{D}_+^*(x-x_1,t)\dot{D}_+(x'-x_1',t')\langle E_m^{x_1} E_n^{x_1'}\rangle +$$

$$+[D_+^*(x-x_1,t)\dot{D}_+(x'-x_1',t') + \frac{1}{c}\dot{D}_+^*(x-x_1,t)D_+(x'-x_1',t')]\times$$

$$\times[\langle E_m^{x_1} Z_n^{x_1'}\rangle + 2i\pi\hbar c(\delta_{mn}\Delta_1 - \frac{\partial^2}{\partial x_{1m}\partial x_{1n}})\delta(x_1 - x_1')] \Big\} .$$

Correlation function $G_{nl}^{(2,2)}(y_1 y_2, y_1' y_2')$ can be calculated only approximately. For example, for the field-plasma system the formula

$$G_{mn}^{(1,1)}(y_1,y_2;y_1',y_2') = G_{m_1 n_1}^{(1,1)}(y_1,y_1')G_{m_2 n_2}^{(1,1)}(y_2,y_2') + G_{m_1 n_2}^{(1,1)}(y_1,y_2')G_{m_2 n_1}^{(1,1)}(y_2,y_1') + O(\lambda^1) . \tag{81}$$

is obtained.

So, the method of the reduced description of nonequilibrium states allows calculating Glauber correlation functions in important models. It gives possibility to analyze correlation properties of electromagnetic field interacting with emitters and plasma in the considered examples. Such analysis can be performed in terms of average electromagnetic field and binary correlations of the field.

Quantum theory of radiation transfer is an important part of quantum optics (Perina, 1984). The problem is: to choose parameters that describe radiation transfer in a medium and obtain a closed set of equations for such parameters. This problem can be solved in the reduced description method.

In the theory of radiation transfer (Chandrasekhar, 1950) energy fluxes in medium and polarization of the radiation are problems of interest. Operator of energy flux is given by the formula

$$\hat{q}_n(x) = \frac{c}{8\pi}\varepsilon_{nlm}\{\hat{E}_l(x), \hat{B}_m(x)\} . \tag{82}$$

In the developed above theory average values of binary in the field quantities can be calculated exactly. For the field-plasma model the following result can be obtained in terms of the one-particle density matrix and Wigner distribution function

$$q_n(x) = \frac{\hbar c^2}{V}\sum_{kq,\alpha\alpha'} n^{\alpha\alpha'}_{k-q/2,k+q/2}\varphi^{\alpha\alpha'}_n(k,q)e^{iqx} = \frac{\hbar c^2}{V}\sum_{k,\alpha\alpha'}\varphi^{\alpha\alpha'}_n(k,-i\frac{\partial}{\partial x})\mathrm{f}^{\alpha\alpha'}_k(x) \tag{83}$$

where

$$\varphi^{\alpha\alpha'}_n(k,q) \equiv \varphi^{\alpha\alpha'}_n(k-q/2, k+q/2), \tag{84}$$

$$\varphi^{\alpha_1\alpha_2}_n(k_1,k_2) = \frac{1}{2}(\delta_{nl}\delta_{ms} - \delta_{ml}\delta_{ns})(k_1k_2)^{1/2}\{\tilde{k}_{1l}e^*_{\alpha_1k_1s}e_{\alpha_2k_2m} + \tilde{k}_{2l}e^*_{\alpha_1k_1m}e_{\alpha_2k_2s}\} .$$

For a weakly nonuniform states of the field formula (82) can be simplified and gives (at $V\to\infty$) a classic expression

$$q_n(x) = \sum_{\alpha}\int\frac{d^3k}{(2\pi)^3}\omega_k\hbar\frac{\partial\omega_k}{\partial k_n}\mathrm{f}^{\alpha\alpha}_k(x) . \tag{85}$$

Formula (83) should be put in the basis of the theory of radiation transfer. The simplest consideration is based on the approximate expression (85). Radiation transfer can be described with specific intensity of radiation in the form

$$I^{\alpha\alpha'}_\omega(n,x) \equiv \frac{\omega^3\hbar}{(2\pi)^3c^2}\mathrm{f}^{\alpha\alpha'}_k(x)\Big|_{k=n\frac{\omega}{c}} \qquad (|n|=1) \tag{86}$$

Therefore, an equation of radiation transfer can be based on the kinetic equation for the Wigner distribution function of the field. According to definition (43) and equation (68), for weakly nonuniform states in the absence of the average field this kinetic equation is written as follows

$$\partial_t\mathrm{f}^{\alpha\alpha'}_k = -\frac{\partial\Omega_k}{\partial k_n}\frac{\partial\mathrm{f}^{\alpha\alpha'}_k}{\partial x_n} - 2\nu_k(\mathrm{f}^{\alpha\alpha'}_k - n_k\delta_{kk'}\delta_{\alpha\alpha'}) + \frac{1}{4}\frac{\partial^2\nu_k}{\partial k_n\partial k_l}\frac{\partial^2\mathrm{f}^{\alpha\alpha'}_k}{\partial x_n\partial x_l} . \tag{87}$$

The radiation transfer equation follows from the definition (86) and kinetic equation (87)

$$\partial_t I^{\alpha\alpha'}_\omega(n,x) = -c_\omega n_l\frac{\partial I^{\alpha\alpha'}_\omega(n,x)}{\partial x_l} - 2\nu_\omega\{I^{\alpha\alpha'}_\omega(n,x) - I_\omega\delta_{\alpha\alpha'}\} + \{a_\omega n_l n_m + b_\omega\delta_{lm}\}\frac{\partial^2 I^{\alpha\alpha'}_\omega(n,x)}{\partial x_l\partial x_m} , \tag{88}$$

where the notations

$$\frac{\partial\tilde{\omega}_k}{\partial k_l}\Big|_{k=\frac{\omega}{c}n} \equiv c_\omega n_l , \qquad \frac{\partial^2\nu_k}{\partial k_l\partial k_m}\Big|_{k=\frac{\omega}{c}n} \equiv a_\omega n_l n_m + b_\omega\delta_{lm} , \qquad \nu_k\big|_{k=\frac{\omega}{c}n} \equiv \nu_\omega ; \tag{89}$$

$$I_\omega \equiv \frac{\omega^3 \hbar}{(2\pi)^3 c^2} \frac{1}{e^{\hbar\omega/T} - 1}$$

are introduced. Usually this equation is written for stationary states and given without correction with the last term. So, the reduced description method provides an approach in which it is possible to justify the radiation transfer theory.

In quantum optics functional methods are widely used. Starting point of such methods is a definition of a generating functional (3) for average values calculated with considered statistical operator ρ. This functional gives possibility of calculating all necessary average values

$$\mathrm{Sp}\rho c^{+}_{\alpha_1 k_1} \ldots c^{+}_{\alpha_s k_s} c_{\alpha'_1 k'_1} \ldots c_{\alpha'_{s'} k'_{s'}} = (-1)^{s'} \frac{\partial^{s+s'} F(u,u^*)}{\partial u_{\alpha_1 k_1} \ldots \partial u_{\alpha_s k_s} \partial u^*_{\alpha'_1 k'_1} \ldots \partial u^*_{\alpha'_{s'} k'_{s'}}} \Bigg|_{u,u^*=0} . \tag{90}$$

Hence, the generating functional gives complete description of a system and evolution equation for this functional is equivalent to the quantum Liouville equation. Definition (3) shows that the functional obeys the property

$$F(u,u^*)^* = F(-u,-u^*) \cdot \tag{91}$$

Let us suppose that effective photon interaction in a system has the form

$$\hat{H} = \sum_1 \varepsilon_1 c_1^+ c_1 + \sum_{123} \{\Phi(12,3) c_1^+ c_2^+ c_3 + h.c.\} \tag{92}$$

where notations

$$c_i = c_{\alpha_i k_i}, \quad c_i^+ = c^+_{\alpha_i k_i}, \quad \varepsilon_i = \omega_{k_i} c, \quad \Phi(12,3) = \Phi(\alpha_1 k_1, \alpha_2 k_2; \alpha_3 k_3), \quad \sum_i = \sum_{\alpha_i k_i} \tag{93}$$

are introduced. The following evolution equation for $F(u,u^*,t)$ can be easily obtained analogously to (Akhiezer & Peletminskii, 1981) from the Liouville equation

$$i\hbar \partial_t F(u,u^*,t) = \sum_1 \varepsilon_1 \Big(u_1 \frac{\partial}{\partial u_1} - u_1^* \frac{\partial}{\partial u_1^*}\Big) F(u,u^*,t) + \tag{94}$$

$$+ \sum_{123} \left\{ \Phi(12,3) \left(\Big(u_3 - \frac{\partial}{\partial u_3^*}\Big) \frac{\partial^2}{\partial u_1 \partial u_2} + \Big(u_1^* - \frac{\partial}{\partial u_1}\Big)\Big(u_2^* - \frac{\partial}{\partial u_2}\Big) \frac{\partial}{\partial u_3^*} \right) + c.c. \right\} F(u,u^*,t)$$

Instead of the generating functional the Glauber-Sudarshan distribution (Glauber, 1969; Klauder & Sudarshan, 1968)

$$P(z,z^*) = \frac{1}{\pi} \int d^2 u F(u,u^*) e^{\sum_{\alpha k} (u^*_{\alpha k} z_{\alpha k} - u_{\alpha k} z^*_{\alpha k})}, \quad F(u,u^*) = \int d^2 z P(z,z^*) e^{\sum_{\alpha k} (u_{\alpha k} z^*_{\alpha k} - u^*_{\alpha k} z_{\alpha k})} \tag{95}$$

is widely used. Formula (95) shows that this distribution is the Fourier transformed generating functional. Note that an evolution equation for the Glauber-Sudarshan distribution can be easily obtained by substituting the second formula in (95) into equation (94). Such evolution equations can be a starting point for constructing the reduced description of a system (Peletminskii, S. & Yatsenko A., 1970). Obtaining the field evolution picture in terms of P-function is very attractive from the point of view of analysis of field properties under consideration in quantum optics.

8. Conclusions

Kinetic theory of electromagnetic field in media has choosing a set of parameters describing nonequilibrium states of the field as a starting point with necessity. The minimal set of such parameters includes binary correlations of field amplitudes. The corresponding mathematical apparatus uses different structures of averages: one-particle density matrices, Wigner distribution functions, and conventional simultaneous correlation functions of field operators. All approaches can be connected with each other due to the possibility of expressing the main correlation parameters in various forms. The reduced description method elucidates the construction of kinetic equations in electrodynamics of continuous media (field-plasma, field-emitters systems) and radiation transfer theory. Electromagnetic field properties are discussed in quantum optics in terms of Glauber correlation functions measured in experiments. Theoretical calculation of such functions requires information about the statistical operator of the system under investigation. In the framework of the reduced description method we have succeeded in obtaining the statistical operator of the field in the form that is convenient for calculations in a number of interesting cases.

9. Acknowledgement

The work was partially supported by the State Foundation for Fundamental Research of Ukraine (project No. 25.2/102).

10. References

Akhiezer, A. & Peletminskii, S. (1981). *Methods of Statistical Physics*, Pergamon Press, New York, USA

Akhiezer, A. & Berstetsky, V. (1969). *Quantum Electrodynamics*, Nauka, Moscow, USSR (in Russian)

Banfi, G. & Bonifacio, R. (1975). Superfluorescence and Cooperative Frequency Shift. *Physical Review*, Vol.A12, (November 1975), pp. 2068-2082

Bogolyubov, N. (1946). *Problems of Dynamical Theory in Statistical Physics*, Gostekhizdat, Moscow-Leningrad, USSR (in Russian)

Bogolyubov, N. (Jr.) & Shumovsky, A. (1987). *Superradiance*, JINR publication P17-87-176, Dubna, USSR (in Russian)

Bogolyubov, N. (Jr.), Kozierowski M., Trang, Q. & Shumovsky A. (1988). New Effects in Quantum Electrodynamics. *Physics of Elementary Particles and Atomic Nuclei*, Vol.19, Issue 4, (August 1988), pp. 831-863, ISSN 0367-2026 (in Russian)

Chandrasekhar S. (1950). *Radiative Transfer*, Oxford, UK

De Groot S. & Suttorp L. (1972). *Fundamentals of Electrodynamics*, North Holland Publishing Company, Amsterdam, Netherlands

Dicke, R. (1954). Coherence in Spontaneous Radiation Processes. *Physical Review*, Vol.93, No.1, (January 1954), pp. 99-110

Glauber, R. (1963). The Quantum Theory of Optical Coherence. *Physical Review*, Vol.130, Issue 6, (June 1963), pp. 2529-2539

Glauber R. (1965). Optical Coherence and Photon Statistics (Les Houches, 1964), In: *Quantum Optics and Electronics*, C. DeWitt, A. Blandin, C. Cohen-Tannoudji (Eds.), pp. 63-107, Gordon and Breach, New York, USA

Glauber, R. (1969). Coherence and Quantum Detection, In: *Rendiconti della Scuola Internazionale di Fisica Enrico Fermi: XLII Corso: Quantum Optics,* BN 708757, Academic Press, New York, London

Hanbury Brown, R. & Twiss, R. (1956). A Test of a New Type of a Stellar Interferometer on Sirius. *Nature,* Vol.178, No.4541, pp. 1046-1048.

Kadantseva, E., Lyagushyn, S. & Shumovsky. A. (1989). *Superradiation with Losses,* Preprint JINR P17-89-431, Dubna, USSR (in Russian)

Kilin, S. (2003). *Quantum Optics. Fields and their Detection,* Editorial URSS, ISBN 5-354-00442-X, Moscow, Russia (in Russian)

Klauder, J. & Sudarshan, E. (1968). *Fundamentals of Quantum Optics,* W.A. Benjamin, Inc., New York, Amsterdam

Landau, L. & Lifshitz E. (1988). *Theoretical Physics (in 10 volumes). Vol.2: Field Theory,* Nauka, Moscow, USSR (in Russian)

Louisell W. (1964). *Radiation and Noise in Quantum Electronics,* McGrow-Hill, New York, USA

Lyagushyn, S., Salyuk, Y. & Sokolovsky A. (2005). Kinetics of Emitters of Electromagnetic Field in the Bogolyubov Reduced Description Method. *Physics of Particles and Nuclei,* Vol.36, Issue Suppl. 1, (November 2005), pp. S123-S127, ISSN 1063-7796

Lyagushyn, S., Salyuk, Y. & Sokolovsky A. (2008). Correlation Properties of Electromagnetic Field Generated by Emitters with Random Orientation. *Proceedings of 12th International Conference Of Mathematical Methods in Electromagnetic Theory,* ISBN 978-1-4244-2284-5, Odesa, Ukraine, June-July 2008

Lyagushyn, S. & Sokolovsky A. (2010). Electromagnetic Field State Description with Correlation Functions of Different Nature. *Proceedings of 2010 International Conference Of Mathematical Methods in Electromagnetic Theory,* ISBN 978-1-4244-8860-5, Kyiv, Ukraine, September 2010

Lyagushyn, S. & Sokolovsky A. (2010). Kinetics of System of Emitters and Nonequilibrium Electromagnetic Field. *Physics of Particles and Nuclei,* Vol.41, Issue 7, (December 2010), pp. 1035-1038, ISSN 1063-7796

Lyagushyn, S., Salyuk, Y. & Sokolovsky A. (2011). Observable Quantities in Kinetics of Electromagnetic Field in Medium. *Proceedings of International School –Seminar " New Physics and Quantum Chromodynamics at External Conditions,* Dnipropetrovs'k, Ukraine, May 2011

Peletminskii S., Prikhod'ko V. & Shchelokov V. (1975). Behavior of Electrodynamic Green's Functions at Low Frequencies. *Theoretical and Mathematical Physics,* Vol.25, No.1, (October 1975), pp. 986-991, ISSN 0040-5779

Peletminskii S. & Yatsenko A. (1970). Generating Functional Method and Virial Expansions in Nonequilibrium Statistical Mechanics. *Theoretical and Mathematical Physics,* Vol.3, No.2, (May 1970), pp. 513-523, ISSN 0040-5779

Perina J. (1984). *Quantum Statisticcs* of *Linear and Nonlinear Optical Phenomena,* D. Reidel Publishing Company, Dodrecht–Boston–Lancaster

Scully, M. & Zubairy, M. (1997). *Quantum Optics,* Cambridge University Press, ISBN 0-521-43458, Cambridge, UK

Sokolovsky, A. & Stupka, A. (2004). Description of Electromagnetic Field in Medium with Strengths and One-Photon Density Matrix. *The Journal of Kharkiv National University, Physical series: Nuclei, Particles, Fields,* Vol.642, Issue 3 (25), (December 2004), pp. 97-100, ISSN 0453-8048 (in Ukrainian)

Sokolovsky, A. & Stupka, A. (2005). Equations of electrodynamics in a hydrodynamic medium with regard for nonequilibrium fluctuations. *Ukrainian Mathematical Journal,* Vol.57, No.6, (June 2005), pp. 1004-1019, ISSN 0041-5995

2

Nonclassical Properties of Superpositions of Coherent and Squeezed States for Electromagnetic Fields in Time-Varying Media

Jeong Ryeol Choi
Department of Radiologic Technology, Daegu Health College,
Yeongsong-ro 15, Buk-gu, Daegu 702-722
Republic of Korea

1. Introduction

In spite of remarkable advance of quantum optics, there would be many things that are yet to be developed regarding the properties of light. One of them is the behavior of light propagating or confined in time-varying media. If the characteristic parameters of medium such as electric permittivity, magnetic permeability, and electric conductivity are dependent on time, the medium is classified as time-varying media. After the publication of a seminal paper by Choi and Yeon (Choi & Yeon, 2005), there has been a surge of renewed research for electromagnetic field quantization in time-varying media and for the properties of corresponding quantized fields (Budko, 2009; Choi, 2010a; Choi, 2010b). Some important examples that the theory of optical wave propagation in time-varying media is applicable are magnetoelastic delay lines (Rezende & Morgenthaler, 1969), wave propagation in ionized plasmas (Kozaki, 1978), the modulation of microwave power (Morgenthaler, 1958), and novel imaging algorithms for dynamical processes in time-varying physical systems (Budko, 2009).

To study the time behavior of light rigorously, it may be crucial to quantize it. The purpose of this chapter is to analyze nonclassical properties of superpositions of quantum states for electromagnetic fields in time-varying linear media. The methods for quantization of a light propagating in free space or in transparent material is well known, since each mode of the field in that case acts like a simple harmonic oscillator. However, the quantization procedure for a light in a time-varying background medium is somewhat complicate and requires elaborate technic in accompanying mathematical treatments. One of the methods that enable us to quantize fields in such situation is to introduce an invariant operator theory (Lewis & Riesenfeld, 1969) in quantum optics. The invariant operator theory which employs Lewis-Riesenfeld invariants is very useful in deriving quantum solutions for time-dependent Hamiltonian systems in cases like this. The light in homogeneous conducting linear media which have time-dependent parameters will be quantized and their quantum properties will be investigated on the basis of invariant operator theory. The exact wave functions for the system with time-varying parameters will be derived in Fock, coherent, and squeezed states in turn.

For several decades, much attention has been devoted to the problem of superposed quantum states (the Schrödinger cat states) of an optical field (Choi & Yeon, 2008; Ourjoumtsev et al., 2006; Yurke & Stoler, 1986). The superpositions in both coherent states and squeezed states of electromagnetic field are proved to be quite interesting and their generation has been an important topic in quantum optics thanks to their nonclassical properties such as high-order squeezing, subpoissonian photon statistics, and oscillations in the photon-number distribution (Richter & Vogel, 2002; Schleich et al., 1991). Moreover, it is shown that the Schrödinger cat states provide an essential tool for quantum information processing (Ourjoumtsev et al, 2006).

It may be interesting to study a phase space distribution function so-called Wigner distribution function (WDF) (Wigner, 1932) for Schrödinger cat states for fields in time-varying media. The propagation of a signal through optical systems is well described by means of the WDF transformations (Bastiaans, 1991), which results in accompaniment of the reconstruction of the propagated signal. A convolution of the WDF allows us to know the phase space distribution connected to a simultaneous measurement of position and momentum. Due to its square integrable property, the WDF always exists and can be employed to evaluate averages of Hermitian observables that are essential in the quantum mechanical theory. The WDF is regarded as 'quasiprobability distribution function', since it can be negative as well as positive on subregions of phase space. Gaussian is the only pure state for which the WDF is positive everywhere. In view of quantum optics, Bastiaans showed that the WDF provides a link between Fourier optics and the geometrical optics (Bastiaans, 1980). The WDF has been widely used in explaining intrinsic quantum features which have no classical analogue in various branches of physics, such as decoherence (Zurek, 1991), Fourier quantum optics (Bartelt et al, 1980), and interference of quantum amplitudes (Bužek et al., 1992). The nonclassical properties of superpositions of quantum states for electromagnetic fields with time-dependent parameters will be studied here via WDF.

2. Quantization of light in time-varying media

The characteristics of electromagnetic fields in media are determined in general by the parameters of media such as electric permittivity ϵ, magnetic permeability μ, and electric conductivity σ. If $\sigma = 0$ and other two parameters are real constants, the electromagnetic fields behave like simple harmonic oscillators. The electromagnetic fields propagating along a medium that have non-zero conductivity undergo dissipation that entails their energy loss. In case that the value of one or more parameters of media is complex and/or time-dependent, the mathematical description of optical fields may be not an easy task. We suppose that the parameters are time-dependent and use invariant operator theory to quantize the electromagnetic fields in such medium. The relations between fields and current in linear media are

$$\mathbf{D} = \epsilon(t)\mathbf{E}, \qquad \mathbf{H} = \frac{\mathbf{B}}{\mu(t)}, \qquad \mathbf{J} = \sigma(t)\mathbf{E}. \tag{1}$$

The Maxwell's equations in media that have no charge source can be written in SI unit as

$$\nabla \cdot \mathbf{D} = 0, \tag{2}$$

$$\nabla \cdot \mathbf{B} = 0, \tag{3}$$

$$\nabla \times \mathbf{E} = -\frac{\partial \mathbf{B}}{\partial t}, \tag{4}$$

$$\nabla \times \mathbf{H} = \mathbf{J} + \frac{\partial \mathbf{D}}{\partial t}. \tag{5}$$

A fundamental relation between electromagnetic fields and potentials are

$$\mathbf{E} = -\nabla\Phi - \frac{\partial \mathbf{A}}{\partial t}, \tag{6}$$

$$\mathbf{B} = \nabla \times \mathbf{A}, \tag{7}$$

where Φ is a scalar potential and $\mathbf{A}$ is a vector potential. We take Coulomb gauge due to its usefulness in this situation. In particular Coulomb gauge is more advantageous in describing a purely transverse wave. The scalar potential then vanishes since we assumed that there is no net charge source. As a consequence, both the electric and the magnetic fields can be expanded only in terms of vector potential.

By solving Eqs. (2)-(5) considering Eqs. (6) and (7), we obtain a time-dependent damped wave equation such that

$$\nabla^2\mathbf{A} - [\sigma(t) + \dot{\epsilon}(t)]\mu(t)\frac{\partial \mathbf{A}}{\partial t} - \epsilon(t)\mu(t)\frac{\partial^2 \mathbf{A}}{\partial t^2} = 0. \tag{8}$$

To decouple the vector potential into position and time functions, it is necessary to put

$$\mathbf{A}(\mathbf{r}, t) = \sum_l \mathbf{u}_l(\mathbf{r})q_l(t), \tag{9}$$

where particular modes are denoted by subscript l. The substitution of Eq. (9) into Eq. (8) leads to

$$\nabla^2\mathbf{u}_l(\mathbf{r}) + k_l^2\mathbf{u}_l(\mathbf{r}) = 0, \tag{10}$$

$$\frac{\partial^2 q_l(t)}{\partial t^2} + \frac{\sigma(t) + \dot{\epsilon}(t)}{\epsilon(t)}\frac{\partial q_l(t)}{\partial t} + c^2(t)k_l^2 q_l(t) = 0, \tag{11}$$

where k_l are separation constants and $c(t)$ is the time-dependent velocity of light which is given by $c(t) = 1/\sqrt{\epsilon(t)\mu(t)}$. Actually k_l are wave numbers that can be represented as

$$k_l = \frac{\omega_l(t)}{c(t)}, \tag{12}$$

where $\omega_l(t)$ are time-dependent natural angular frequencies. From the fact that k_l are constants, we have

$$\frac{\omega_l(t)}{c(t)} = \frac{\omega_l(0)}{c(0)}. \tag{13}$$

From now on, let us omit under subscript l from notations for the shake of convenience.

Using fundamental theory of dynamics, we can construct the Hamiltonian of the system associated with Eq. (11) to be

$$\hat{H}(\hat{q}, \hat{p}, t) = \frac{1}{2\epsilon_0}e^{-\Lambda(t)}\hat{p}^2 + \frac{1}{2}e^{\Lambda(t)}\epsilon_0\omega^2(t)\hat{q}^2, \tag{14}$$

where $\epsilon_0 = \epsilon(0)$ and $\Lambda(t) = \int_0^t \{[\sigma(t') + \dot{\epsilon}(t')]/\epsilon(t')\}dt'$. If we consider that this Hamiltonian is a time-varying form, the introduction of a suitable invariant operator $\hat{K}$ may enable us to obtain quantum solutions of the system. The invariant operator can be evaluated from

$$\frac{d\hat{K}}{dt} = \frac{\partial \hat{K}}{\partial t} + \frac{1}{i\hbar}[\hat{K}, \hat{H}] = 0, \tag{15}$$

which is known as Liouville-von Neumann equation. Execution of some algebra after inserting Eq. (14) into the above equation gives

$$\hat{K} = \left(\frac{\Omega}{2\rho(t)}\hat{q}\right)^2 + [\rho(t)\hat{p} - \epsilon_0 e^{\Lambda(t)}\dot{\rho}(t)\hat{q}]^2, \tag{16}$$

where Ω is an arbitrary real positive constant and $\rho(t)$ is some real time-function that satisfies the following differential equation

$$\ddot{\rho}(t) + \frac{\sigma(t) + \dot{\epsilon}(t)}{\epsilon(t)}\dot{\rho}(t) + \omega^2(t)\rho(t) - \frac{\Omega^2}{4\epsilon_0^2}e^{-2\Lambda(t)}\frac{1}{\rho^3(t)} = 0. \tag{17}$$

If we introduce annihilation and creation operators of the form

$$\hat{a} = \sqrt{\frac{1}{\hbar\Omega}}\left[\left(\frac{\Omega}{2\rho(t)} - i\epsilon_0 e^{\Lambda(t)}\dot{\rho}(t)\right)\hat{q} + i\rho(t)\hat{p}\right], \tag{18}$$

$$\hat{a}^\dagger = \sqrt{\frac{1}{\hbar\Omega}}\left[\left(\frac{\Omega}{2\rho(t)} + i\epsilon_0 e^{\Lambda(t)}\dot{\rho}(t)\right)\hat{q} - i\rho(t)\hat{p}\right], \tag{19}$$

the invariant operator can be rewritten as

$$\hat{K} = \hbar\Omega\left(\hat{a}^\dagger\hat{a} + \frac{1}{2}\right). \tag{20}$$

Note that Eqs. (18) and (19) are different from those of simple harmonic oscillator.

If we denote two linearly independent homogeneous solutions of Eq. (11) as $\rho_1(t)$ and $\rho_2(t)$, $\rho(t)$ is given by (Eliezer & Gray, 1976)

$$\rho(t) = [h_1\rho_1^2(t) + h_2\rho_1(t)\rho_2(t) + h_3\rho_2^2(t)]^{1/2}, \tag{21}$$

where h_1, h_2, and h_3 are constants that follow some relation imposed between them. In terms of an Wronskian w which is a time-constant and has the form

$$\mathsf{w} = \epsilon_0 e^{\Lambda(t)}[\rho_1(t)\dot{\rho}_2(t) - \dot{\rho}_1(t)\rho_2(t)], \tag{22}$$

h_1, h_2, and h_3 yield

$$4h_1h_3 - h_2^2 = \Omega^2/\mathsf{w}^2. \tag{23}$$

As an example, we can take the electromagnetic parameters to be

$$\epsilon(t) = \epsilon_0 e^{\gamma t}, \qquad \mu(t) = \mu(0), \qquad \sigma(t) = 0, \tag{24}$$

where γ is a real constant. Then, the Wronskian can be rewritten as

$$\mathsf{w} = \epsilon(t)[\rho_1(t)\dot{\rho}_2(t) - \dot{\rho}_1(t)\rho_2(t)], \tag{25}$$

where $\rho_1(t)$ and $\rho_2(t)$ are given by

$$\rho_1(t) = \rho_{1,0}\, e^{-\gamma t/2} J_1[\xi(t)], \tag{26}$$

$$\rho_2(t) = \rho_{2,0}\, e^{-\gamma t/2} N_1[\xi(t)]. \tag{27}$$

Here, $\rho_{1,0}$ and $\rho_{2,0}$ are arbitrary real constants and $\xi(t) = [2\omega(0)/\gamma]e^{-\gamma t/2}$.

If we consider the asymptotic behavior of Bessel functions for $x \gg 1$:

$$J_m(x) \simeq \sqrt{\frac{2}{\pi x}} \cos\left(x - m\frac{\pi}{2} - \frac{\pi}{4}\right), \tag{28}$$

$$N_m(x) \simeq \sqrt{\frac{2}{\pi x}} \sin\left(x - m\frac{\pi}{2} - \frac{\pi}{4}\right), \tag{29}$$

$\rho_1(t)$ and $\rho_2(t)$, in the limit $\xi \gg 1$ with a selection of $\rho_{1,0} = -\rho_{2,0} = \sqrt{\pi\Omega/(2\epsilon_0\gamma)}$, becomes

$$\rho_1(t) \simeq \sqrt{\frac{\Omega}{\epsilon_0\gamma\xi(t)}} e^{-\gamma t/2} \cos\left(\xi(t) - \frac{3\pi}{4}\right), \tag{30}$$

$$\rho_2(t) \simeq -\sqrt{\frac{\Omega}{\epsilon_0\gamma\xi(t)}} e^{-\gamma t/2} \sin\left(\xi(t) - \frac{3\pi}{4}\right). \tag{31}$$

Then, $\Omega/(2\mathrm{w}) = 1$ and, as a consequence, we can choose $h_1 = h_3 = 1$ and $h_2 = 0$ so that Eq. (21) reduces to $\rho(t) = [\rho_1^2(t) + \rho_2^2(t)]^{1/2}$ which is a well used relation in the literature (Choi, 2010b).

We can directly check that the ladder operators satisfy the boson commutation relation $[\hat{a}, \hat{a}^\dagger] = 1$. Therefore, it is possible to obtain zero-point eigenstate $\langle q|\phi_0(t)\rangle$ of $\hat{K}$ from $\hat{a}\langle q|\phi_0(t)\rangle = 0$ and nth order eigenstate by operating $\hat{a}^\dagger$ n times into $\langle q|\phi_0(t)\rangle$. Thus we finally get

$$\langle q|\phi_n(t)\rangle = \sqrt[4]{\frac{\Omega}{2\rho^2(t)\hbar\pi}} \frac{1}{\sqrt{2^n n!}} H_n\left(\sqrt{\frac{\Omega}{2\rho^2(t)\hbar}} q\right) \times \exp\left[-\frac{1}{2\rho(t)\hbar}\left(\frac{\Omega}{2\rho(t)} - i\epsilon_0 e^{\Lambda(t)}\dot{\rho}(t)\right) q^2\right], \tag{32}$$

where H_n is nth order Hermite polynomial. For the time-dependent Hamiltonian systems in cases like this, the Schrödinger solutions are different from the eigenstates $\langle q|\phi_n(t)\rangle$ by only time-dependent phase factors (Lewis & Riesenfeld, 1969):

$$\langle q|\psi_n(t)\rangle = \langle q|\phi_n(t)\rangle \exp[i\theta_n(t)]. \tag{33}$$

If we insert this equation together with Eq. (14) into Schrödinger equation, we obtain the phases $\theta_n(t)$ as:

$$\theta_n(t) = -\left(n + \frac{1}{2}\right) \frac{\Omega}{2\epsilon_0} \int_0^t \frac{e^{-\Lambda(t')}}{\rho^2(t')} dt'. \tag{34}$$

The probability of finding the real photons is given by probability density that is squared modulus of the wave function, while the wave function itself in general has no physical reality.

The probability density $|\langle q|\psi_n(t)\rangle|^2$ in number state is illustrated in Fig. 1 as a function

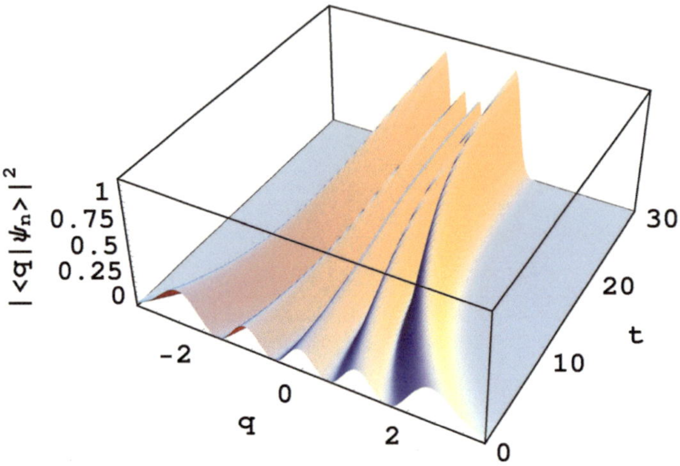

Fig. 1. Probability density in number state as a function of q and t. We used $n = 4$, $\hbar = 1$, $\gamma = 0.3$, $\epsilon_0 = 1$, $\mu(0) = 1$, $\Omega = 1$, $c_1 = c_2 = 25$, $\omega(0) = 1$, $h_1 = h_3 = \Omega/(2\text{w})$, $h_2 = 0$, and $\rho_{1,0} = \rho_{2,0} = 0.5$. All values are taken to be dimensionless for convenience (This convention will be used in all figures in this chapter).

of q and t. The parameters chosen in this figure are the same as those of Eq. (24). (This choice will also be hold in all subsequent figures without mentioning.) The probability density converges to origin ($q = 0$) as time goes by. This means that the amplitude of electromagnetic wave decreases with time. In general, the damping factor $(\sigma + \dot{\epsilon})/\epsilon$ appeared in Eq. (11) is responsible for the dissipation of amplitude. As you can see form Eq. (24), it is kept that $\sigma = 0$ in our model example but $\dot{\epsilon}$ is not zero. Therefore, the variation of $\epsilon(t)$ with time is the actual factor that leads to take place the dissipation of amplitude in this case. Note that $\epsilon(t)$ exponentially increases depending on γ. Thus,for large γ, the amplitude decreases more rapidly.

The density operator ϱ is defined in the form

$$\hat{\varrho} = \sum_{n,m} \varrho_{nm} |\psi_n\rangle\langle\psi_m|. \tag{35}$$

Then, the WDF is represented in terms of $\hat{\varrho}$ as

$$W(q,p,t) = \frac{1}{\pi\hbar}\int_{-\infty}^{\infty} \langle q - x|\hat{\varrho}|q + x\rangle e^{2ipx/\hbar} dx. \tag{36}$$

The properties of superposition states are well understood from WDF representation. A little algebra gives

$$W(q,p,t) = \frac{1}{\pi\hbar}\int_{-\infty}^{\infty} \hat{\varrho}(q - x, q + x, t) e^{2ipx/\hbar} dx. \tag{37}$$

It is well known that the WDF for number state consists of many concentric circles of ridge and valley which have different radii (Choi, 2004). The total number of ridge and valley is associated with the quantum number of the system and the value of WDF at a valley is negative. Whenever the WDF takes on negative values in parts of the phase, the corresponding state is regarded as nonclassical one. Due to its allowance of negative values, it is impossible to interpret WDF as the real distribution function. For this reason, there have been established several kinds of weighted WDF that takes non-negative value in phase space (Mogilevtsev & Kilin, 2000). The negative values of WDF have indeed been observed from lots of experimental measurements for a variety of states of optical field and matter (Leonhardt, 1997). However, integration of WDF over either of q and p makes it to be probability distribution for the other

$$\int_{-\infty}^{\infty} W(q,p,t)dq = |\psi(p,t)|^2, \tag{38}$$

$$\int_{-\infty}^{\infty} W(q,p,t)dp = |\psi(q,t)|^2. \tag{39}$$

These formulae guarantee WDF to be a quantum distribution function (but quasi) in spite of its singular properties.

3. Superposition of coherent states

The state engineering can also be achieved using a coherent-state expansion, instead of expanding number-state whose wave function is derived in the previous section. Coherent state for harmonic oscillator were firstly found by Schrödinger (Schrödinger, 1926) and rediscovered afterwards by Glauber (Glauber, 1963). Though coherent states are classical-like quantum states and hardly exhibit nonclassical effects, most class of superposition of coherent states can exhibit one or more nonclassical effects among various possible nonclassicality such as sub-Poissonian photon statistics and squeezing.

The coherent state $|\alpha\rangle$ is an eigenstate of the annihilation operator:

$$\hat{a}|\alpha\rangle = \alpha|\alpha\rangle. \tag{40}$$

If we consider Eq. (18), α is given by

$$\alpha = \sqrt{\frac{1}{\hbar\Omega}}\left[\left(\frac{\Omega}{2\rho(t)} - i\epsilon_0 e^{\Lambda(t)}\dot{\rho}(t)\right) q_{\rm cl}(t) + i\rho(t)p_{\rm cl}(t)\right], \tag{41}$$

where $q_{\rm cl}(t)$ and $p_{\rm cl}(t)$ are classical trajectories of variables q and p, which are given by

$$q_{\rm cl}(t) = c_1\rho_1(t) + c_2\rho_2(t), \tag{42}$$

$$p_{\rm cl}(t) = \epsilon_0 e^{\Lambda(t)}\frac{dq_{\rm cl}(t)}{dt} = \epsilon_0 e^{\Lambda(t)}[c_1\dot{\rho}_1(t) + c_2\dot{\rho}_2(t)], \tag{43}$$

where c_1 and c_2 are arbitrary real constants. If we divide α into real and imaginary parts such that

$$\alpha_{\rm R} = \sqrt{\frac{\Omega}{\hbar}}\frac{1}{2\rho(t)}[c_1\rho_1(t) + c_2\rho_2(t)], \tag{44}$$

$$\alpha_{\rm I} = \sqrt{\frac{1}{\hbar\Omega}}\epsilon_0 e^{\Lambda(t)}\{\rho(t)[c_1\dot{\rho}_1(t) + c_2\dot{\rho}_2(t)] - \dot{\rho}(t)[c_1\rho_1(t) + c_2\rho_2(t)]\}, \tag{45}$$

the eigenstate can be represented in terms of amplitude α_0 and phase φ:

$$\alpha = \alpha_0 e^{i\varphi}, \tag{46}$$

where

$$\alpha_0 = \sqrt{\alpha_{\rm R}^2 + \alpha_{\rm I}^2}, \tag{47}$$

$$\varphi = \tan^{-1}(\alpha_{\rm I}/\alpha_{\rm R}). \tag{48}$$

The substitution of Eqs. (44) and (45) into Eqs. (47) and (48) leads to

$$\alpha_0 = \sqrt{\frac{\Omega(c_1^2 h_3 - c_1c_2h_2 + c_2^2 h_1)}{\hbar(4h_1h_3 - h_2^2)}}, \tag{49}$$

$$\varphi(t) = \tan^{-1}\left(\frac{2h_1c_2\rho_1(t) - h_2[c_1\rho_1(t) - c_2\rho_2(t)] - 2h_3c_1\rho_2(t)}{\sqrt{4h_1h_3 - h_2^2}[c_1\rho_1(t) + c_2\rho_2(t)]}\right). \tag{50}$$

The time behavior of $\varphi(t)$ is illustrated in Fig. 2. The considered domain for $\varphi(t)$ in this figure is $-\pi/2 < \varphi(t) < \pi/2$, i.e., $\varphi(t) \equiv m\pi + \delta(t) \rightarrow \delta(t)$ where m is an integer and $-\pi/2 < \delta(t) < \pi/2$ at a given time. The direct differentiation of Eq. (50) with respect to time gives

$$\frac{d\varphi(t)}{dt} = -\frac{\Omega e^{-\Lambda(t)}}{2\epsilon_0\rho^2(t)}. \tag{51}$$

Thus, we can represent φ in another way such that

$$\varphi(t) = -\frac{\Omega}{2\epsilon_0}\int_0^t \frac{e^{-\Lambda(t')}}{\rho^2(t')}dt' + \varphi(0). \tag{52}$$

It may be instructive to compare this equation with Eq. (34). The time behavior of $\varphi(t)$ is the same as that of $\theta_n(t)$ when we neglect some constants.

By operating $\langle q|$ from left in Eq. (40), the coherent state in configuration space is obtained. Then, a suitable choice of phase leads to (Choi & Yeon, 2008)

$$\langle q|\alpha\rangle = \sqrt[4]{\frac{\Omega}{2\rho^2(t)\hbar\pi}}\exp\left[\alpha\sqrt{\frac{\Omega}{\rho^2(t)\hbar}}q - \frac{1}{4\rho(t)\hbar}\right.$$
$$\left.\times\left(\frac{\Omega}{\rho(t)} - 2i\epsilon_0 e^{\Lambda(t)}\dot{\rho}(t)\right)q^2 - \frac{1}{2}\alpha_0^2 - \frac{1}{2}\alpha^2\right]. \tag{53}$$

The relation between coherent state and number state eigenfunction is given by

$$\langle q|\alpha\rangle = \exp\left(-\frac{1}{2}\alpha_0^2\right)\sum_n \frac{\alpha^n}{\sqrt{n!}}\langle q|\phi_n(t)\rangle. \tag{54}$$

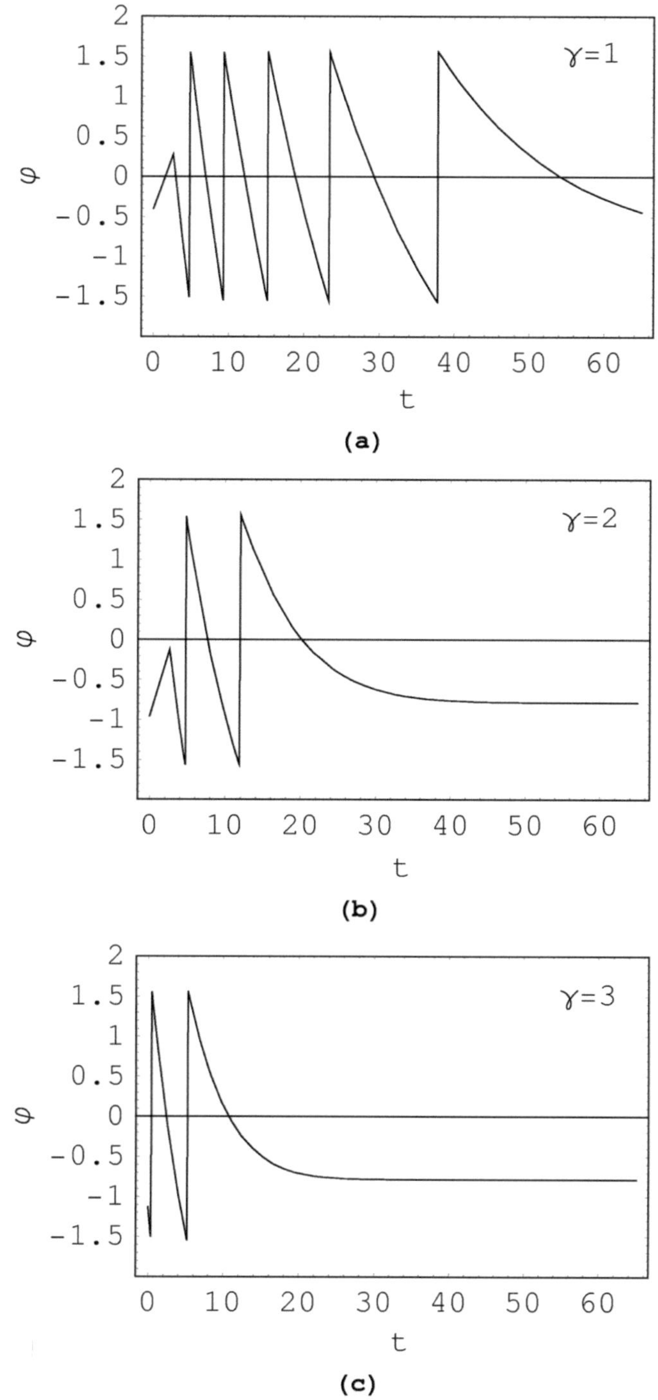

Fig. 2. The time evolution of φ for various values of γ. We used $\epsilon_0 = 1$, $\Omega = 1$, $c_1 = c_2 = 25$, $\omega(0) = 1$, $h_1 = h_3 = \Omega/(2\mathrm{w})$, $h_2 = 0$, and $\rho_{1,0} = \rho_{2,0} = 0.5$.

We can easily show that the probability density $|\langle q|\alpha\rangle|^2$ is Gaussian from a fundamental evaluation. As mentioned earlier, the only pure states for which the WDF is positive everywhere are those that their corresponding probability density is Gaussian in cases like this. We can confirm from Fig. 3 that the trajectory of the peak of $|\langle q|\alpha\rangle|^2$ oscillates like a

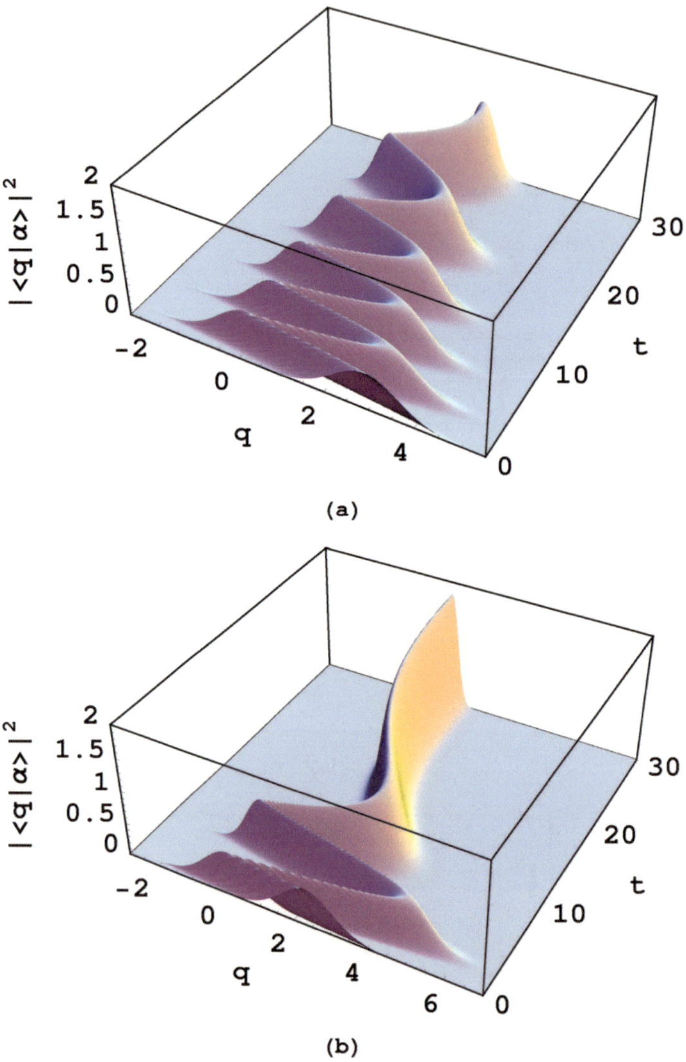

Fig. 3. Probability density in coherent state as a function of q and t. The value of γ is 0.1 for (a) and 0.3 for (b). We used $\hbar = 1$, $\epsilon_0 = 1$, $\mu(0) = 1$, $\Omega = 1$, $c_1 = c_2 = 25$, $\omega(0) = 1$, $h_1 = h_3 = \Omega/(2\mathrm{w})$, $h_2 = 0$, and $\rho_{1,0} = \rho_{2,0} = 0.5$.

classical state and converges near to origin as time goes by due to the influence of damping factor γ. Although coherent state is a pure quantum state, its properties lie on a borderline between those of classical state and quantum states.

From the early days of quantum mechanics, there have been great efforts for the problem of generating arbitrary quantum states including nonclassical states of an optical field mode. In particular, the superposition of coherent state (which is the main topic in this section) and the superposition of squeezed state (that will be treated in the next section) attracted much attention in the literature. A widely accepted criterion that a state to be classified as nonclassical one is exist: A quantum state has nonclassicality when the Glauber-Sudarshan P-function (Glauber, 1963; Sudarshan, 1963) fails to show the properties of a classical probability density. However, in many cases, this definition may hardly be applied to investigate the nonclassicality for a direct interpretation of experiments due to quite singular characteristics of P-function and the difficulty in determining P-function from given measurements. In fact, even for the simple harmonic oscillator, the exact characterization of the nonclassicality of a quantum state in terms of measurable quantities is somewhat ambiguous. A hierarchy of observable conditions for nonclassical quantum states, which allows one to verify whether the P-function for a specific state shows the properties of a classical probability density or not, has been reported (Richter & Vogel, 2002), while the global criteria for nonclassicality of states are yet the subject of researches.

Meanwhile, a method to reconstruct characteristic functions of a quantum state, such as the density matrix, the WDF, and the P-function, from experimentally accessible data is established, which is known as optical homodyne tomography (Smithey et al., 1993; Kiesel et al., 2008). It is possible to reconstruct the P-function up to sufficiently large thermal photon number whereas other criteria for nonclassicality, such as the Klyshko criterion (Klyshko, 1996), negativities of the WDF, and the entanglement potential (Asbóth, 2005), start to fail as the number of thermal photon increases (Kiesel et al., 2008). Though both definitions of nonclassicality in terms of P-function and in terms of WDF are sufficient but not necessary and leave some families of nonclassical quantum states outside their scope (Lvovsky & Shapiro, 2002). Of course, a satisfaction of the requirements of either definition does not automatically grantee satisfaction of the other. While the condition for nonclassicality based on P-function is more general than that based on WDF and covers more broad range of nonclassical quantum states, the negativity of the WDF, that is our main concern in this Chapter, is recognized as very strong indication for nonclassical character of quantum states.

The nonclassical properties of quantized light is highlighted by superposing two distinct states. Let us consider a superposition of two coherent states, that the corresponding wave function is represented in the form

$$\langle q|\psi(t)\rangle = \frac{1}{\sqrt{N}}(\langle q|\alpha_0 e^{i\varphi}\rangle + e^{i\phi}\langle q|\alpha_0 e^{-i\varphi}\rangle), \tag{55}$$

where

$$N = 2\{1 + \exp(-2\alpha_0^2 \sin^2\varphi)\cos[\alpha_0^2 \sin(2\varphi) - \phi]\}. \tag{56}$$

Here, the total phase difference between two constituent states in superposition is 2φ, and the relative phase between the two components of the superposition is ϕ. Strictly speaking, this definition of cat state is somewhat different from that of Tara et al (Tara et al., 1993) [or that of Schleich et al. (Schleich et al., 1991), for $\phi = 0$]: The cat state of Tara et al. is defined in terms of $|\alpha e^{i\vartheta}\rangle$ and $|\alpha e^{-i\vartheta}\rangle$ instead of $|\alpha_0 e^{i\varphi}\rangle$ and $|\alpha_0 e^{-i\varphi}\rangle$, where ϑ is an arbitrary real constant. The interaction of coherent states with nonlinear medium can be a source for generating superposed coherent states (Tara et al, 1993). Not only the quadrature squeezing but also the Sub-Poissonian and oscillatory photon statistics are typical consequences of nonclassical effects of quantum interference produced by superposition.

While the coherent states among all pure quantum states have the most properties of classicality, their superposition represented in Eq. (55) reveals remarkable features of nonclassicality. By substituting Eq. (53) into Eq. (55), we easily get the wave function in configuration space:

$$\langle q|\psi(t)\rangle = \sqrt[4]{\frac{\Omega}{2\rho^2(t)\hbar\pi}}\frac{2}{\sqrt{N}}\exp\left[-\frac{1}{4\rho(t)\hbar}\left(\frac{\Omega}{\rho(t)} - 2i\epsilon_0 e^{\Lambda(t)}\dot{\rho}(t)\right)q^2 - \frac{1}{2}\alpha_0^2 + \frac{i\phi}{2}\right]$$
$$\times\exp\left(\alpha_0\sqrt{\frac{\Omega}{\rho^2(t)\hbar}}q\cos\varphi - \frac{1}{2}\alpha_0^2\cos(2\varphi)\right)$$
$$\times\cos\left(\alpha_0\sqrt{\frac{\Omega}{\rho^2(t)\hbar}}q\sin\varphi - \frac{1}{2}\alpha_0^2\sin(2\varphi) - \frac{\phi}{2}\right). \quad (57)$$

Though the illustration for the probability density $|\langle q|\psi(t)\rangle|^2$ given in Fig. 4 is somewhat complicate, the principal trajectory of $|\langle q|\psi(t)\rangle|^2$ is very similar to that of $|\langle q|\alpha\rangle\rangle|^2$ given in Fig. 3. The superposition in a case like this become a family of Schrödinger cat states only when the amplitude of the electromagnetic field is sufficiently large. Theoretical results of several previous researches (Kis et al., 2001; Varada &. Agarwal, 1993) show that certain quantum states can be approximated by superposing macroscopically distinguishable coherent states.

As is well known, WDF provides a possible method to describing a quantum system in terms of a quasi-distribution in phase space. It enables us to analyze the interference between two component states involved in the superposition. The WDF for the superposed coherent state is obtained from

$$W(q,p,t) = \frac{1}{\pi\hbar}\int_{-\infty}^{\infty}\langle\psi(t)|q+x\rangle\langle q-x|\psi(t)\rangle e^{2ipx/\hbar}dx. \quad (58)$$

Performing the integration after inserting Eq. (57) into the above equation gives

$$W(q,p,t) = \frac{2}{\pi\hbar N}\exp\left(2\sqrt{\frac{\Omega}{\hbar\rho^2(t)}}q\alpha_0\cos\varphi\right)\exp\left(-\frac{2}{\hbar\Omega}K(q,p,t)\right)$$
$$\times[\exp(-2\alpha_0^2)\cosh\Theta_1 + \exp(-2\alpha_0^2\cos^2\varphi)\cos(\Theta_2 - \phi)], \quad (59)$$

where

$$K(q,p,t) = \frac{\Omega^2}{4\rho^2(t)}q^2 + \left[\rho(t)p - \epsilon_0 e^{\Lambda(t)}\dot{\rho}(t)q\right]^2, \quad (60)$$

$$\Theta_1 = \frac{4\rho(t)\alpha_0\sin\varphi}{\sqrt{\Omega\hbar}}\left(p - \epsilon_0 e^{\Lambda(t)}\frac{\dot{\rho}(t)}{\rho(t)}q\right), \quad (61)$$

$$\Theta_2 = 2\sqrt{\frac{\Omega}{\hbar\rho^2(t)}}q\alpha_0\sin\varphi - \alpha_0^2\sin(2\varphi). \quad (62)$$

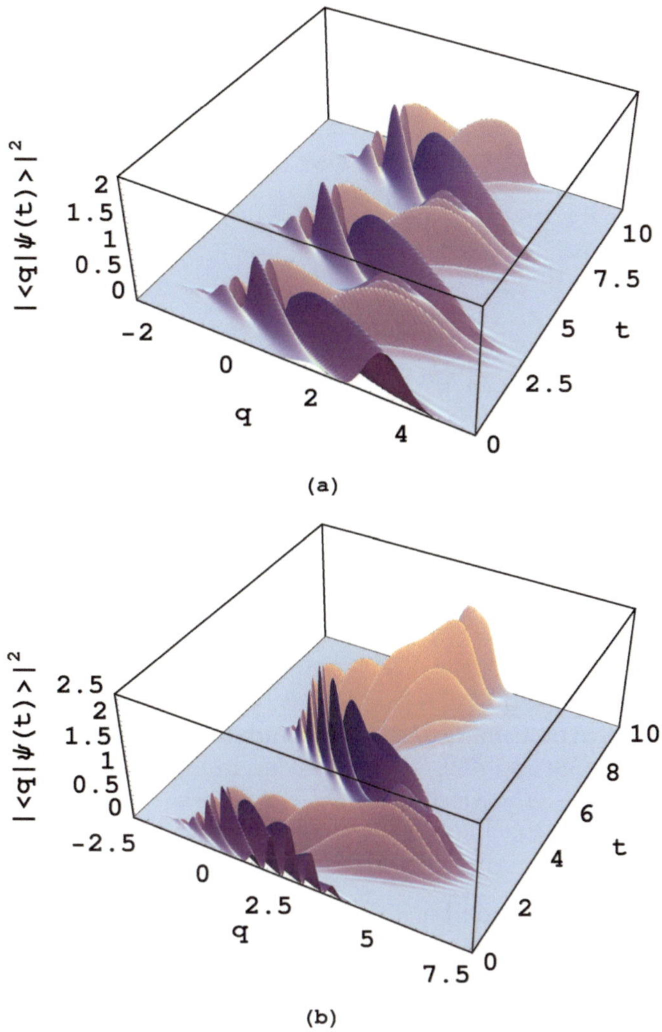

Fig. 4. Probability density for superposition of coherent states as a function of q and t. The value of γ is 0.1 for (a) and 0.3 for (b). We used $\hbar = 1, \epsilon_0 = 1, \mu(0) = 1, \phi = 1, \Omega = 1,$ $c_1 = c_2 = 25, \omega(0) = 1, h_1 = h_3 = \Omega/(2\mathrm{w}), h_2 = 0,$ and $\rho_{1,0} = \rho_{2,0} = 0.5$.

From Fig. 5, we can find the nonclassical characteristics of the superposed coherent state. The two bells that are Gaussian type correspond to the two constituent coherent states, and the ripple given in the middle between them is taken place from quantum interference. Interference occurs when the two bells do not overlap, but the WDF should have non-zero values along their common intervals in q and/or p coordinates (Dragoman, 2001). As you can see, some parts of the ripple take on negative value. This is a clear signal for the existence of nonclassical features of the system. The number of peaks in the structure of the ripple becomes large as the distance between the two bells increases. We can confirm from this aspect that the wavelength of interference fringe is inversely proportional to the value of α_0. The wavelength

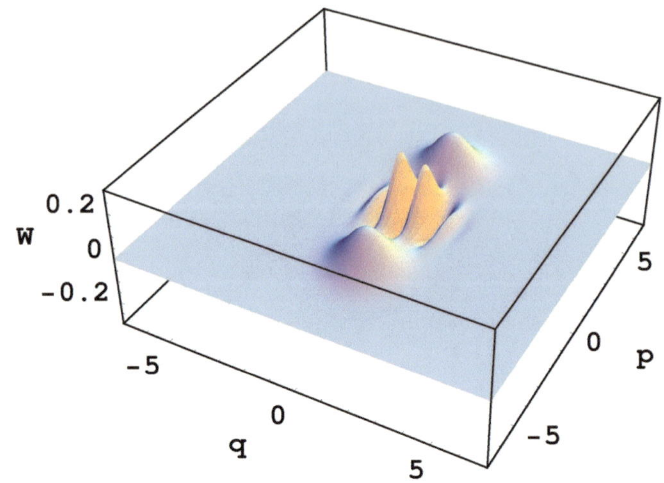

Fig. 5. Quadrature plot of WDF for superposition of coherent states. We used $\hbar = 1, \gamma = 0.1$, $\epsilon_0 = 1, \mu(0) = 1, \phi = 1, \Omega = 1, c_1 = c_2 = 25, \omega(0) = 1, t = 1, h_1 = h_3 = \Omega/(2\mathrm{w}), h_2 = 0$, and $\rho_{1,0} = \rho_{2,0} = 0.5$.

of interference fringe is a major factor that determine the shape of interference in probability density displayed in Fig. 4 (and Figs. 6 and 7 for the superposed squeezed state). Recall that the interference pattern in probability density for an arbitrary superposition state is connected with its WDF via Eqs. (38) and (39). If we consider, in the context of classical mechanics, that the physical attributes of a system exist objectively even when it is unknown, classical mechanics fails to give a reasonable explanation for the negative values in the superposition states.

If we consider that the last term in Eq. (59) involving $\cos(\Theta_2 - \phi)$ is the interference term, the structure of interference varies according to the value of ϕ. The superposition with $\phi = 0$ for simple harmonic oscillator and the corresponding WDF are studied in detail by several researchers (Bužek et al., 1992; Raimond et al., 1997; Schleich et al., 1991; Varada &. Agarwal, 1993). In particular, some researchers (Simon et al., 1997; Yurke & Stoler, 1986) are interested in superpositions with $\phi = \pi/2$ for a little different aspect than here, thanks to their experimental realizability, for this family of states, through the evolution of a coherent state in a Kerr medium. For $\phi = \pi/2$, Eq. (59) becomes

$$W(\phi = \pi/2; q, p, t) = \frac{2}{\pi\hbar N_{\pi/2}} \exp\left[-2\sqrt{\frac{\Omega}{\hbar\rho^2(t)}} q\alpha_0 \cos\varphi\right] \exp\left(-\frac{2}{\hbar\Omega}K(q,p,t)\right)$$
$$\times[\exp(-2\alpha_0^2)\cosh\Theta_1 + \exp(-2\alpha_0^2\cos^2\varphi)\sin\Theta_2], \quad (63)$$

where

$$N_{\pi/2} = 2\{1 + \exp(-2\alpha_0^2 \sin^2\varphi)\sin[\alpha_0^2\sin(2\varphi)]\}. \quad (64)$$

Nonclassical features for this family of states are studied extensively in the literature (Ahmad et al., 2011).

4. Superposition of squeezed states

The investigation of the properties of squeezed state and its generation is also a central topic in quantum optics since it enables us to utilize an optical field with reduced quadrature noise. We introduce a squeeze operator as

$$\hat{b} = \mu\hat{a} + \nu\hat{a}^{\dagger}, \tag{65}$$

where μ and ν obey

$$|\mu|^2 - |\nu|^2 = 1. \tag{66}$$

As is the case of $\hat{a}$ and $\hat{a}^{\dagger}$, this operator and its Hermitian conjugate satisfy the boson commutation relation, $[\hat{b}, \hat{b}^{\dagger}] = 1$. Let us consider only real values for μ and ν on the purpose to simplify the problem. Squeezed state $|\beta\rangle$ is the eigenstate of $\hat{b}$:

$$\hat{b}|\beta\rangle = \beta|\beta\rangle. \tag{67}$$

For convenience in further study, we introduce a squeezing parameter as $d = \mu/\nu$. Then, the wave function of squeezed state in configuration space can be evaluated in terms of d using Eq. (67):

$$\langle q|\beta\rangle = \mathcal{N}_q \exp\left[-\frac{q^2}{2\rho(t)\hbar}\left(\frac{\Omega}{2\rho(t)}\frac{d+1}{d-1} - i\epsilon_0 e^{\Lambda(t)}\dot{\rho}(t)\right) + \frac{d\alpha + \alpha^*}{d-1}\sqrt{\frac{\Omega}{\hbar\rho^2(t)}}q\right], \tag{68}$$

where a normalization factor $\mathcal{N}_q$ is given by

$$\mathcal{N}_q = \left(\frac{\Omega}{2\rho^2(t)\hbar\pi}\frac{d+1}{d-1}\right)^{1/4} \exp\left(-\frac{d+1}{d-1}\alpha_0^2\cos^2\varphi + i\delta_{s,q}(\alpha,\alpha^*)\right), \tag{69}$$

with an arbitrary real phase $\delta_{s,q}(\alpha,\alpha^*)$. Considering Eqs. (66) and (67), it is easy to show that the eigenvalue β can be written in the form

$$\beta = \mu\alpha + \nu\alpha^*. \tag{70}$$

Now we represent β as

$$\beta = \beta_0 e^{i\varphi_\beta}, \tag{71}$$

where β_0 and φ_β. are real. Execution of an algebra with the substitution of Eq. (46) into Eq. (70) yields

$$\beta_0 = \alpha_0\sqrt{\mu^2 + \nu^2 + 2\mu\nu\cos(2\varphi)}, \tag{72}$$

$$\varphi_\beta = \tan^{-1}\left(\frac{d-1}{d+1}\tan\varphi\right). \tag{73}$$

Let us take $\delta_{s,q}(\alpha,\alpha^*)$ in the form

$$\delta_{s,q}(\alpha,\alpha^*) = -\alpha_0^2\sin\varphi\cos\varphi. \tag{74}$$

Then, Eq. (68) reduces to a simple form which is

$$\langle q|\beta\rangle = \left(\frac{\Omega}{2\pi\hbar\rho^2(t)}\frac{d+1}{d-1}\right)^{1/4} \exp\left[-\frac{1}{2\rho(t)\hbar}\left(\frac{\Omega}{2\rho(t)}\frac{d+1}{d-1} - i\epsilon_0 e^{\Lambda(t)}\dot{\rho}(t)\right)q^2 \right.$$
$$\left. + \frac{d\alpha+\alpha^*}{d-1}\left(\sqrt{\frac{\Omega}{\hbar\rho^2(t)}}q - \alpha_0\cos\varphi\right)\right]. \tag{75}$$

The squeezed state which have this wave function belongs to a nonclassical state. If we recall that any non-commuting observables in quantum mechanics can be determined simultaneously in classical mechanics with any order of precision, classical analogue for squeezing is unthinkable.

A superposition of squeezed states may also be useful in understanding nonclassical features of quantum states. Let us consider the superposition of $\langle q|\beta\rangle$ and $\langle q|\beta^*\rangle$ with an arbitrary relative phase ϕ. The wave function for this system is given by

$$\langle q|\Psi(t)\rangle = \frac{1}{\sqrt{\mathcal{N}}}(\langle q|\beta_0 e^{i\varphi_\beta}\rangle + e^{i\phi}\langle q|\beta_0 e^{-i\varphi_\beta}\rangle), \tag{76}$$

where $\mathcal{N}$ is a normalization constant of the form

$$\mathcal{N} = 2\left[1+\exp\left(-\frac{2(d-1)}{d+1}\alpha_0^2\sin^2\varphi\right)\cos[\alpha_0^2\sin(2\varphi)-\phi]\right]. \tag{77}$$

Substituting Eq. (75) into Eq. (76) and, then, executing some algebra results in

$$\langle q|\Psi(t)\rangle = \left(\frac{\Omega}{2\rho^2(t)\hbar\pi}\frac{d+1}{d-1}\right)^{1/4}\frac{2}{\sqrt{\mathcal{N}}}\exp\left[-\frac{1}{2\rho(t)\hbar}\left(\frac{\Omega}{2\rho(t)}\frac{d+1}{d-1} - i\epsilon_0 e^{\Lambda(t)}\dot{\rho}(t)\right)q^2 \right.$$
$$\left. + \frac{(d+1)\alpha_0\cos\varphi}{d-1}\left(\sqrt{\frac{\Omega}{\hbar\rho^2(t)}}q - \alpha_0\cos\varphi\right) + \frac{i\phi}{2}\right]$$
$$\times\cos\left[\alpha_0\sin\varphi\left(\sqrt{\frac{\Omega}{\hbar\rho^2(t)}}q - \alpha_0\cos\varphi\right) - \frac{\phi}{2}\right]. \tag{78}$$

Figures 6 and 7 are probability densities, $|\langle q|\Psi(t)\rangle|^2$, with the squeezing for q-quadrature and for p-quadrature, respectively. By comparing these with Fig. 4, we see that the width of densities are narrowed for the squeezing for q-quadrature and broadened for p-quadrature. Thus the uncertainty of q is reduced for the case of the squeezing for q-quadrature, while increased for p-quadrature. Therefore, through application of squeezed states, we are able to reduce noise dispersion in one quadrature at the expense of increased noise in the complementary quadrature as compared with that of coherent state. For this reason, squeezed states of light can be applied to high precision interferometer that provides high resolutions in measurement beyond the standard limits.

All wave functions that are given in Eqs. (32)[or (33)], (53), (57), (75), and (78) satisfy Schrödinger equation when supplemented by some time-dependent phase factors suitable

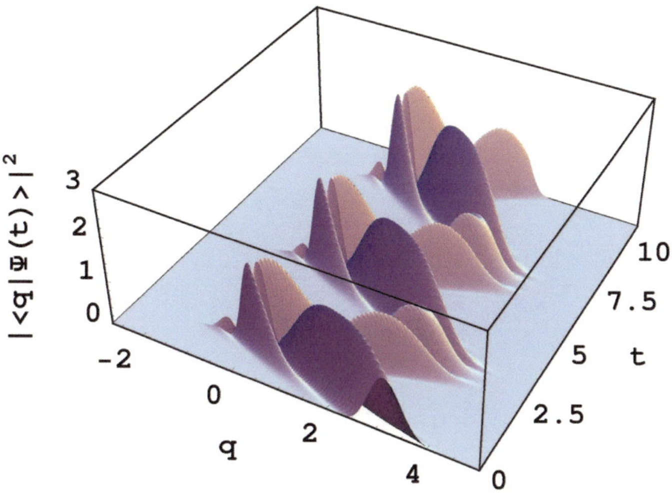

(a)

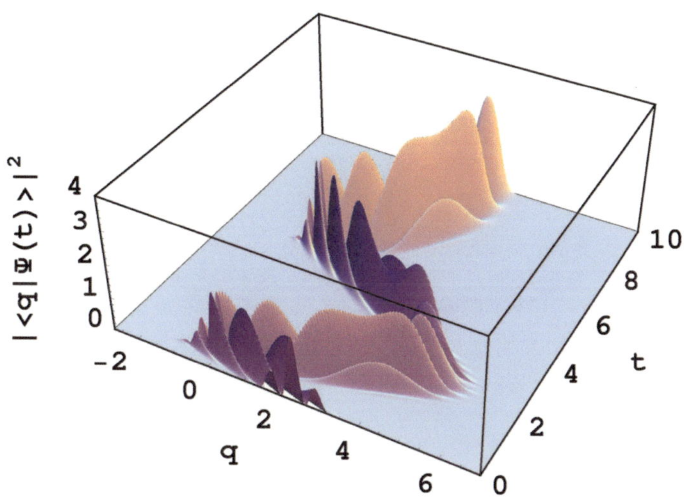

(b)

Fig. 6. Probability density for superposition of squeezed states as a function of q and t. The value of γ is 0.1 for (a) and 0.3 for (b). The same values as that of Fig. 4 are used except for $d = 2$.

for each. The phase (factor) for number state is given in Eq. (34) and that for other wave functions can also be derived from the same method as that of the number state (Choi, 2011).

The quantum wave functions have no reality of their own and are just associated with the probability to find photon in a certain domain as mentioned previously. The mathematical description in classical optics for the interference in phase space is very similar to that in quantum optics, but the classically represented optical waves have a real character.

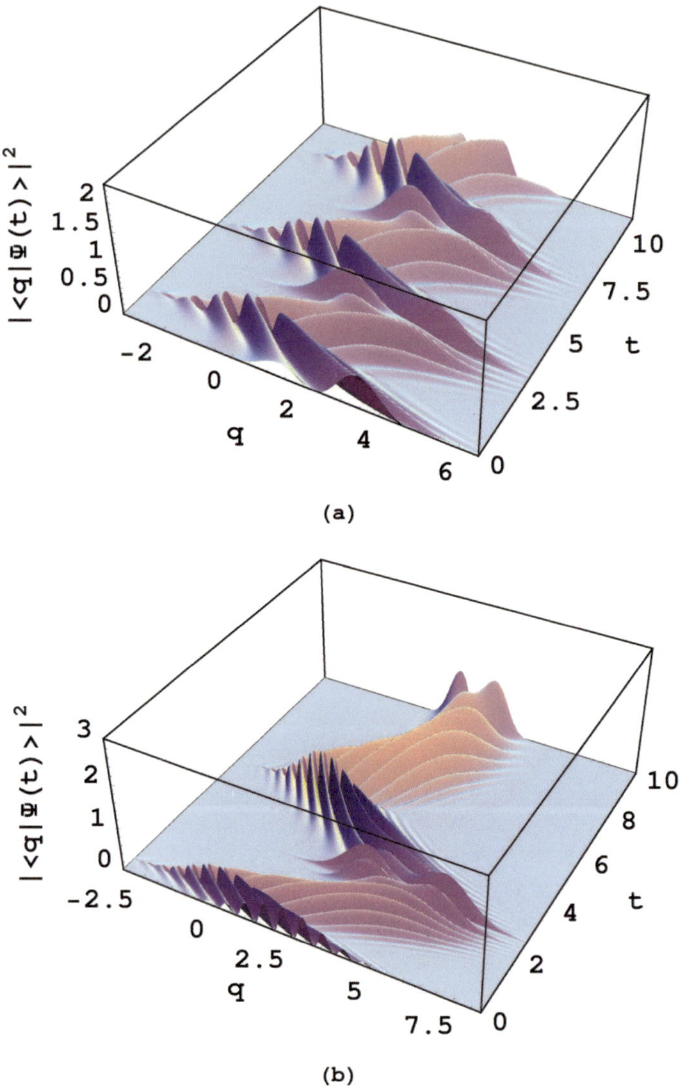

Fig. 7. Probability density for superposition of squeezed states as a function of q and t. The value of γ is 0.1 for (a) and 0.3 for (b). The same values as that of Fig. 4 are used except for $d = -2$.

Using the same method as that of previous section, the WDF that corresponds to Eq. (78) is evaluated to be

$$\mathcal{W}(q,p,t) = \frac{2}{\pi\hbar\mathcal{N}} \exp\left(-\frac{2}{\hbar\Omega}K_s(q,p,t)\right) \times \exp\left[\frac{d+1}{d-1}\left(2\sqrt{\frac{\Omega}{\hbar\rho^2(t)}}q\alpha_0\cos\varphi - 2\alpha_0^2\cos^2\varphi\right)\right] \times \left[\exp\left(-\frac{d-1}{d+1}2\alpha_0^2\sin^2\varphi\right)\cosh\left(\frac{d-1}{d+1}\Theta_1\right) + \cos(\Theta_2 - \phi)\right], \quad (79)$$

where

$$K_s(q,p,t) = \frac{\Omega^2}{4\rho^2(t)} \frac{d+1}{d-1} q^2 + \frac{d-1}{d+1} [\rho(t)p - \epsilon_0 e^{\Lambda(t)} \dot{\rho}(t) q]^2. \tag{80}$$

The WDF with squeezing is plotted in Fig. 8: (a) corresponds to squeezing for q-quadrature

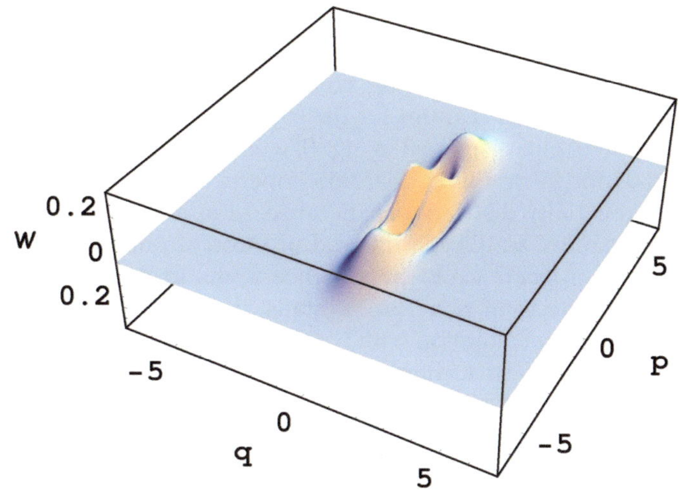

(a)

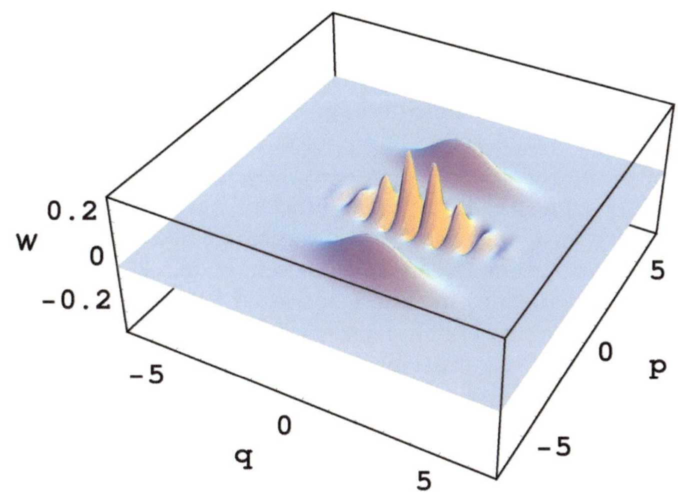

(b)

Fig. 8. Quadrature plot of WDF for superposition of squeezed states with $d = 3$ for (a) and $d = -2$ for (b). All values used here are the same as those in Fig. 5.

and (b) for p-quadrature. The width of two bells is shortened along the direction of q for (a) and shortened along the direction of p for (b) when they are compared to that of coherent state shown in Fig. 5.

We can represent the position of two bells in polar coordinate as $(\alpha_0, \pm\varphi)$ (Schleich et al., 1991). For $\varphi = 0$ and $\varphi = \pi$, the two bells overlap: There are no interferences in this case, but are quantum transitions between two different quantum states overlapped (Dragoman, 2001). On the other hand, for $\varphi = \pi/2$, they are separated from each other with maximum distance. The microscopical pattern of the structure of interference is directly related to the value of ϕ. Of course, these rules mentioned in this paragraph are equally applied to the case of the superposition of coherent state.

The WDF plays a crucial role in analyzing nonclassical characteristics of quantum states. You can confirm the nonclassical features for the superposed squeezed states from negative values in interference structure displayed in Fig. 8, which are very similar to that of previous section. The methods for interpreting quantum superpositions are different from that for simple addition of probability distributions, because, in quantum mechanics, we deal with superpositions of probability amplitudes instead of those of probabilities themselves. This is closely related to the appearance of interference terms in the distribution functions of probability. The novel effects of nonclassical states that admits no analogue in classical mechanics have drawn special attention both in theoretical and experimental physics thanks to their applicability in modern technology employing optical and/or other dynamical systems (Ourjoumtsev et al, 2006).

The development of modern technology in experimental photon engineering have made it possible to produce Schrödinger cat states and/or kitten states (small Schrödinger cat states) on the basis of effective nonlinear operations that can be realized via projective measurements and post-selection. Projective measurements based on the Hilbert space formulation of quantum theory produce complete determinations of the post-measurement states through the projection-valued measures of a Hermitian operator (von Neumann, 1932). Kitten states can be produced by squeezing a single-photon. An interesting and useful way to obtain a squeezed single-photon is subtracting one photon from a single-mode squeezed vacuum beam generated by an optical nonlinear process, so-called degenerate optical parametric down-conversion (Ourjoumtsev et al., 2006). A sufficiently large Schrödinger cat states with a smaller overlap between two constituent states can be created by subtracting multiphoton from a squeezed vacuum beam (Neergaard-Nielsen et al., 2011). Other methods for preparation of superposition states include a squeezed Schrödinger cat state prepared by conditional homodyning of a two-photon Fock state (Ourjoumtsev et al., 2007), high-fidelity superposition states prepared using cavity QED technology (de Queirós et al., 2007), and preparation of entangled non-local superposition states (Ourjoumtsev et al., 2009).

5. Conclusion

Nonclassical features of superpositions of coherent states and squeezed states for electromagnetic field in linear media whose electromagnetic parameters vary with time are examined. The expansion of Maxwell equations in charge-source free medium gives second order differential equations for both position function $\mathbf{u}(\mathbf{r})$ and time function $q(t)$. The Hamiltonian associated with the classical equation of motion for $q(t)$ varies with time. Among several methods that are useful in managing time-dependent Hamiltonian systems, the quantum invariant operator method is used in order to solve quantum solutions of the system. The annihilation and the creation operators related to quantum invariant operator satisfy boson commutation relation. The wave functions in number state are derived by taking advantage of the annihilation and the creation operators.

Coherent state is obtained from the expansion of the wave functions of number state. By solving the eigenvalue equation of squeeze operator, squeezed state is also obtained. We can confirm from Figs. 4, 6 and 7, that the detailed structure of probability densities for superposition states are somewhat complicated due to the interference between the two component states. We cannot observe nonclassical properties of the coherent state from WDF, because the value of WDF for a single coherent state is always positive. However, a minor nonclassicality of the coherent state has been reported (Johansen, 2004): A particular quantum distribution function for the coherent state of simple harmonic oscillator, the so-called Margenau-Hill distribution, can take negative values in some regions, but the negative values are relatively very small. This appearance reflects the nonclassicality demonstrated in weak measurements which are, in general, performed under the situation where the coupling of a measuring device to the measured system is very weak. The average values obtained from the weak measurement reveal a time-symmetric dependence on initial and final conditions (Shikano & Hosoya, 2010), providing a natural definition of conditional probabilities in quantum mechanics and, consequently, enabling a more complete description for quantum statistics. However, the interpretation of the results of weak measurements is somewhat controversial on account of its peculiar feature that the measured (weak) value can take strange ones which are outside the range of the eigenvalues of a target observable and may even be complex. The detailed analysis of the strange features of weak measurements may provide a better understanding for the essential differences between quantum and classical statistics.

From Figs. 5 and 8, we can see the interference structure produced between the two main bells. If we consider Eqs. (38) and (39), this determines the pattern of interference fringe in real space of q and p. Though two main bells are always positive, interference structure takes positive and negative values in turn, where the appearance of the negative values is an important signal for the nonclassicality of the system. In fact, nonclassical quantum states are general and ubiquitous. Not only any pure state of the harmonic oscillator can be represented in terms of nonclassical quantum states but also even the number state is a class of nonclassical quantum states (Kis et al., 2001). Nonclassical states in physical fields such as various optical systems, ion motion in a Paul trap, and quantum dot can be applied to fundamental problems in modern technology ranging from high-resolution spectroscopy to low-noise communication and quantum information processing (Kis et al., 2001). In particular, nonclassical properties of correlated quantum systems are expected to play the key role to overcome some limitations relevant to information processing in classical computer system and classical communication.

6. References

Ahmad, M. A.; Bukhari, S. H.; Khan, S. N.; Ran, Z.; Liao, Q. & Liu, S. (2011). Nonclassical features of entangled coherent states. *J. Mod. Opt.*, Vol. 58, No. 10, pp. 890-895.

Asbóth, J. K.; Calsamiglia, J. & Ritsch, H. (2005). Computable measure of nonclassicality for light. *Phys. Rev. Lett.*, Vol. 94, No. 17, pp. 173602(1-4).

Bartelt, H. O.; Brenner, K.-H. & Lohmann, A. W. (1980). The Wigner distribution function and its optical production. *Opt. Commun.*, Vol. 32, No. 1, pp. 32-38.

Bastiaans, M. J. (1980). The Wigner distribution function and its applications in optics. In: M.A. Machado and L.M. Narducci, (Editors), *Optics in Four Dimensions*, AIP Conf. Proc. Vol. 65, pp. 292-312, American Institute of Physics, New York.

Bastiaans, M. J. (1991). Second-order moments of the Wigner distribution function in first-order optical systems. *Optik*, Vol. 88, No. 3, pp. 163-168.

Budko, N. V. (2009). Electromagnetic radiation in a time-varying background medium. *Phys. Rev. A*, Vol. 80, No. 5, pp. 053817(1-8).

Bužek, V.; Vidiella-Barranco, A. & Knight, P. L. (1992). Superpositions of coherent states: Squeezing and dissipation. *Phys. Rev. A*, Vol. 45, No. 9, pp. 6570-6585.

Choi, J. R. (2004). Wigner distribution for the time-dependent quadratic-Hamiltonian quantum system using the Lewis-Riesenfeld invariant operator. *Int. J. Theore. Phys.*, Vol. 44, No. 3, pp. 327-348.

Choi, J. R. (2010a). Invariant operator theory for the single-photon energy in time-varying media. *Chinese Phys. B* Vol. 19, No. 1, pp. 010306(1-5).

Choi, J. R. (2010b). Interpreting quantum states of electromagnetic field in time-dependent linear media. *Phys. Rev. A*, Vol. 82, No. 5, pp. 055803(1-4).

Choi, J. R. (2011). An approach to dark energy problem through linear invariants. *Chinese. Phys. C*, Vol. 35, No. 3, pp. 233-242.

Choi, J. R. & Yeon, K. H. (2005). Quantum properties of light in linear media with time-dependent parameters by Lewis-Riesenfeld invariant operator method. *Int. J. Mod. Phys. B*, Vol. 19, No. 14, pp. 2213-2224.

Choi, J. R. & Yeon, K. H. (2008). Time-dependent Wigner distribution function employed in coherent Schrodinger cat states: $|\Psi(t)\rangle = N^{-1/2}(|\alpha\rangle + e^{i\phi}|-\alpha\rangle)$. *Phys. Scr.*, Vol. 78, No. 4, pp. 045001(1-9).

de Queirós, I. P.; Cardoso, W. B. & de Almeida, N. G. (2007). Superposition of coherent states prepared in one mode of a dissipative bimodal cavity. *J. Phys. B: At. Mol. Opt. Phys.*, Vol. 40, No. 1, pp. 21-27.

Dragoman, D. (2001). Quantum interference as phase space filtering. *Optik*, Vol. 112, No. 1, pp. 31-36.

Eliezer, C. J. & Gray, A. (1976). A note on time-dependent harmonic-oscillator. *SIAM J. Appl. Math.*, Vol. 30, No. 3, pp. 463-468.

Glauber, R. J. (1963). Coherent and incoherent states of the radiation field. *Phys. Rev.*, Vol. 131, No. 6, pp. 2766-2788.

Johansen, L. M. (2004). Nonclassical properties of coherent states. *Phys. Lett. A*, Vol. 329, No. 3, pp. 184-187.

Kiesel, T.; Vogel, W.; Parigi, V.; Zavatta, A. & Bellini, M. (2008). Experimental determination of a nonclassical Glauber-Sudarshan P function. *Phys. Rev. A*, Vol. 78, No. 2, pp. 021804(R)(1-4).

Kis, Z.; Vogel, W. & Davidovich, L. (2001). Nonlinear coherent states of trapped-atom motion. *Phys. Rev. A*, Vol. 64, No. 3, pp. 033401(1-10).

Klyshko, D. N. (1996). Observable signs of nonclassical light. *Phys. Lett. A*, Vol. 213, No. 1-2, pp. 7-15.

Kozaki, S. (1978). Reflection of electromagnetic wave from a time-varying medium. *Electron. Lett.*, Vol. 14, No. 25, pp. 826-828.

Leonhardt, U. (1997). *Measuring the Quantum State of Light*. Cambridge University Press, Cambridge, England.

Lewis, H. R. Jr. & Riesenfeld, W. B. (1969). An exact quantum theory of the time-dependent harmonic oscillator and of a charged particle in a time-dependent electromagnetic field. *J. Math. Phys*, Vol. 10, No. 8, pp. 1458-1473.

Lvovsky, A. I. & Shapiro, J. H. (2002). Nonclassical character of statistical mixtures of the single-photon and vacuum optical states. *Phys. Rev. A*, Vol. 65, No. 3, pp. 033830(1-6).

Mogilevtsev, D. & Kilin, S. (2000). Why the ąőcoarse-grainingąŕ of Wigner function is always coarse. *Opt. Commun.*, Vol. 178, No. 1-3, pp. 147-150.

Morgenthaler, F. R. (1958). Velocity modulation of electromagnetic waves. *IRE Transactions on Microwave Theory and Techniques*, Vol. 6, No. 2, pp. 167-172.

Neergaard-Nielsen, J. S.; Takeuchi, M.; Wakui, K.; Takahashi, H.; Hayasaka, K.; Takeoka, M. & Sasaki, M. (2011). Photon subtraction from traveling fields - recent experimental demonstrations. *Progress in Informatics*, No. 8, pp. 5-18.

Ourjoumtsev, A.; Ferreyrol, F.; Tualle-Brouri, R. & Grangier, P. (2009). Preparation of non-local superpositions of quasi-classical light states. *Nature Physics*, Vol. 5, No. 3, pp. 189-192.

Ourjoumtsev, A.; Jeong, H.; Tualle-Brouri, R. & Grangier, P. (2007). Generation of optical ąőSchrödinger cats' from photon number states. *Nature*, Vol. 448, No. 7155, pp. 784-786.

Ourjoumtsev, A.; Tualle-Brouri, R.; Laurat, J. & Grangier, P. (2006). Generating optical Schrödinger kittens for quantum information processing *Science*, Vol. 312, No. 5770, pp. 83-86.

Raimond, J. M.; Brune, M. & Haroche, S. (1997). Reversible decoherence of a mesoscopic superposition of field states. *Phys. Rev. Lett.*, Vol. 79, No. 11, pp. 1964-1967.

Rezende, S. M. & Morgenthaler, F. R. (1969). Magnetoelastic waves in time-varying magnetic fields. I. Theory. *J. Appl. Phys.*, Vol. 40, No. 2, pp. 524-536.

Richter, Th. & Vogel, W. (2002). Nonclassicality of quantum states: A hierarchy of observable conditions. *Phys. Rev. Lett.*, Vol. 89, No. 28, pp. 283601(1-4).

Schleich, W.; Pernigo, M. & Kien, F. L. (1991). Nonclassical state from two pseudoclassical states. *Phys. Rev. A*, Vol. 44, No. 3, pp. 2172-2187.

Schrödinger, E. (1926). Der stetige Übergang von der Mikro- zur Makromechanik. *Naturwiss.*, Vol. 14, No. 28, pp. 664-666.

Shikano, Y. & Hosoya, A. (2010). Weak values with decoherence. *J. Phys. A: Math. Theor.*, Vol. 43, No. 2, pp. 025304(1-14).

Simon, R.; Selvadoray, M.; Arvind & Mukunda, N. (1997). Necessary and suffient classicality conditions on photon number distributions. arXiv.quant-ph/9709030.

Smithey, D. T.; Beck, M.; Raymer, M. G. & Faridani, A. (1993). Measurement of the Wigner distribution and the density matrix of a light mode using optical homodyne tomography: Application to squeezed states and the vacuum. *Phys. Rev. Lett.*, Vol. 70, No. 9, pp. 1244-1247.

Sudarshan, E. C. G. (1963). Equivalence of semiclassical and quantum mechanical descriptions of statistical light beams. *Phys. Rev. Lett.*, Vol. 10, No. 7, pp. 277-279.

Tara, K.; Agarwal, G. S. & Chaturvedi, S. (1993). Production of Schrödinger macroscopic quantum-superposition states in a Kerr medium. *Phys. Rev. A*, Vol. 47, No. 6, pp. 5024-5029.

Varada, G. V. &. Agarwal, G. S. (1993). Quantum-statistical properties of a particle in a double-harmonic-oscillator potential: Generation of Schrödinger-cat states. *Phys. Rev. A*, Vol. 48, No. 6, pp. 4062-4067.

von Neumann, J. (1932). *Mathematische Grundlagen der Quantenmechanik*, Springer, Berlin, Germany [English translation: Beyer, R. T. (1955). *Mathematical Foundations of Quantum Mechanics*, Princeton University Press, Princeton, NJ, USA].

Wigner, E. (1932). On the quantum correction for thermodynamic equilibrium. *Phys. Rev.*, Vol. 40, No. 5, pp. 749-759.

Yurke, B. & Stoler, D. (1986). Generating quantum mechanical superpositions of macroscopically distinguishable states via amplitude dispersion. *Phys. Rev. Lett.*, Vol. 57, No. 1, pp. 13-16.

Zurek, W. H. (1991). Decoherence and the transition from quantum to classical. *Phys. Today*, Vol. 44, No. 10, pp. 36-40.

3

Photon Localization Revisited

Peeter Saari
Institute of Physics, University of Tartu
Estonia

1. Introduction

Whilst quantum electrodynamics (QED) underwent an impressive development and reached its maturity in the middle of the last century, one of its basic concepts—the photon wave function in free space—was deprived of such fortune. Although the photon wave function in coordinate representation was introduced already in 1930 by Landau and Peierls, the concept was found to suffer from inherent difficulties that were not overcome during the century —see review (Bialynicki-Birula, 1996). The common explanation presented in textbooks, e.g., (Akhiezer & Berestetskii, 1965; Mandel & Wolf, 1995), may be summed up as follows: (i) no position operator exists for the photon, (ii) while the position wave function may be localized near a space-time point, the measurable quantities like the electromagnetic field vectors, energy, and the photodetection probability remain spread out due to their non-local relation with the position wave function.

However, just before the turn of the century both of these widely-espoused notions were disproved (Bialynicki-Birula, 1998; Hawton, 1999) and in the new century a fresh interest in the photon localization problem seems to have been awakened, see, e.g., (Bialynicki-Birula & Bialynicka-Birula, 2009; Chan et al., 2002; Hawton, 2007; Keller, 2000; 2005), meeting the needs of developments in near-field optics, cavity QED, and quantum computing. Recently, into the study of quantum phenomena in general and photon localization in particular, the so-called localized waves were involved (Belgiorno et al., 2010; Besieris et al., 1994; Ciattoni & Conti, 2007; Jáuregui & Hacyan, 2005; Saari et al., 2005). These belong to the propagation-invariant non-diffracting localized solutions to the linear wave equation—a research subject emerged in the 1980-ies, see the 1st collective monograph on the field (Hernández-Figueroa et al., 2008). Experimental feasibility of some of the localized waves has been demonstrated already (Alexeev et al., 2002; Bowlan et al., 2009; Grunwald et al., 2003; Reivelt & Saari, 2002; Saari et al., 2010; Saari & Reivelt, 1997; Sõnajalg et al., 1997).

(Bialynicki-Birula, 1998) writes that the statement "even when the position wave function is strongly concentrated near the origin, the energy wave function is spread out over space asymptotically like $r^{-7/2}$ " —citation from (Mandel & Wolf, 1995), p. 638—is incorrect and that both wave functions may be strongly concentrated near the origin. He demonstrates, on one hand, that photons can be essentially better localized in space—with an exponential falloff of the photon energy density and the photodetection rates. On the other hand, he establishes—and it is even somewhat startling that nobody has done it earlier—that certain localization restrictions arise out of a mathematical property of the positive frequency solutions which therefore are of a universal character and apply not only to photon states but hold for all particles. More specifically, it has been proven in the Letter (Bialynicki-Birula,

1998) for the case of spherically imploding-exploding one-photon wavepacket that the Paley-Wiener theorem allows even at instants of maximal localization only such asymptotic decrease of the modulus of the wave function with the radial distance r that is *slower* than the linear exponential one, i.e., anything slower than $\sim \exp(-Ar)$, where A is a constant. The latter is what the Paley-Wiener theorem says about a function whose Fourier spectrum contains no negative frequencies. In (Bialynicki-Birula & Bialynicka-Birula, 2009) the same results have been obtained using a set of space-dependent photon creation and annihilation operators without any reference to the mode decomposition. The latter paper shows also that one may sharply localize either electric or magnetic but not both footprints of photons.

The purpose of the chapter is to give an overview and an analysis of the most striking contradictions in the notions of photon localization presented in textbooks and in abovementioned new studies.

The first Section reproduces derivation of main textbook formulas concerning the photon localization problem. In the next Section we demonstrate, following (Bialynicki-Birula, 1998), a superficiality of the common textbook notion. Here we also publish for the first time our generalizations to the exponential localization models found in (Bialynicki-Birula, 1998).

In Section 4 we take a closer look at the localization restrictions caused by the circumstance that the photon wave function is mathematically an analytic signal with respect to time variable, which obeys the Paley-Wiener theorem. Here we present some supporting graphical illustrations as well.

Section 5 reproduces results of our paper (Saari et al., 2005) on two-dimensional localization of photon packets constructed from certain localized wave solutions to the wave equation. A discussion follows why such packets seemingly violate the localization restrictions set by the Paley-Wiener theorem.

2. Common treatment of the problem

In this Section we outline the standard approach to the photon localizability problem following the monograph (Mandel & Wolf, 1995). Only these formulas will be presented that are requisite for introduction, comparison and contrasting with what follows in the next Sections.

A field state containing one photon with wavevector $\mathbf{k}$ and polarization s is given through acting of the photon creation operator $a^+(\mathbf{k}, s)$ on the vacuum state $|vac\rangle$. Any one-photon state which is at least partially localized needs a linear superposition of such plane-wave states of the form

$$|1ph\rangle = \sum_s \int d^3k\, \phi(\mathbf{k}, s) a^+(\mathbf{k}, s)\, |vac\rangle \ , \tag{1}$$

in which $\phi(\mathbf{k}, s)$ is any properly normalized weight function. The vector function

$$\mathbf{\Phi}(\mathbf{r}, t) = \sum_s \int d^3k\, \phi(\mathbf{k}, s) \varepsilon_{\mathbf{k},s} e^{i(\mathbf{kr}-\omega t)} \ , \tag{2}$$

where $\varepsilon_{\mathbf{k},s}$ are (generally complex) unit polarization vectors and the frequency $\omega = c\,|\mathbf{k}|$, then represents the corresponding position space wave function (known as the Landau-Peierls wave function) of the photon in state $|1ph\rangle$. Its modulus squared, integrated over a volume V, gives the probability of locating a particle within the volume V.

However, the vector function defined in Eq. (2) does not transform locally under Lorentz transformations. Moreover, there is no procedure to measure $\mathbf{\Phi}(\mathbf{r},t)$. Measurable may be the quantities like energy or the photoelectric detection probability of the photon. It can be shown that neither of them are localized in the volume into which $\mathbf{\Phi}(\mathbf{r},t)$ has been confined.

The average photon energy is

$$\langle 1ph|\,\hat{H}\,|1ph\rangle = \sum_s \int d^3k\,\hbar\omega\,|\phi(\mathbf{k},s)|^2\,.$$

By introducing the function

$$\mathbf{\Phi}_E(\mathbf{r},t) = \sum_s \int d^3k\,\sqrt{\hbar\omega}\phi(\mathbf{k},s)\varepsilon_{\mathbf{k},s}e^{i(\mathbf{kr}-\omega t)}\,, \qquad (3)$$

cf. Eq. (2), which might be called the energy wave function to distinguish it from $\mathbf{\Phi}(\mathbf{r},t)$, and differs from $\mathbf{\Phi}(\mathbf{r},t)$ only in having the $\mathbf{k}$-dependent factor $\sqrt{\hbar\omega}$ in the expansion, we readily find that

$$\langle 1ph|\,\hat{H}\,|1ph\rangle = \int d^3r\,|\mathbf{\Phi}_E(\mathbf{r},t)|^2\,. \qquad (4)$$

$|\mathbf{\Phi}_E(\mathbf{r},t)|^2$ therefore plays the role of energy density. But this energy density is not locally connected with the photon density $|\mathbf{\Phi}(\mathbf{r},t)|^2$. Indeed, from the Fourier expansions of $\mathbf{\Phi}(\mathbf{r},t)$ and $\mathbf{\Phi}_E(\mathbf{r},t)$ we find with the help of the convolution theorem that they are connected through the spatial convolution

$$\mathbf{\Phi}_E(\mathbf{r},t) = \int d^3r'\,G(\mathbf{r}-\mathbf{r}')\mathbf{\Phi}(\mathbf{r}',t)\,, \qquad (5)$$

where the spatial function $G(...)$ is the three-dimensional Fourier transform of $\sqrt{\hbar\omega}$, or

$$G(\mathbf{r}) = \frac{(\hbar c)^{1/2}}{(2\pi)^3}\int d^3k\,\sqrt{k}e^{i\mathbf{kr}}\,. \qquad (6)$$

Because of the spread associated with $G(\mathbf{r})$, $\mathbf{\Phi}_E(\mathbf{r},t)$ can be non-zero at positions where $\mathbf{\Phi}(\mathbf{r},t)$ is zero.

Following (Mandel & Wolf, 1995) and by introducing an exponential factor with vanishing parameter ε in order to regularize the integral, one gets

$$G(\mathbf{r},\varepsilon) = \frac{(\hbar c)^{1/2}}{(2\pi)^3}\int d^3k\,\sqrt{k}e^{i\mathbf{kr}}e^{-\varepsilon k}\,.$$

This integral can be taken in spherical coordinates

$$\begin{aligned} G(\mathbf{r},\varepsilon) &= \frac{(\hbar c)^{1/2}}{(2\pi)^3}\int_0^\infty dk\,k^2\sqrt{k}e^{-\varepsilon k}\int_0^\pi d\theta\,e^{ikr\cos\theta}\sin\theta\int_0^{2\pi}d\phi \\ &= \frac{(\hbar c)^{1/2}}{2\pi^2}\frac{1}{r}\int_0^\infty dk\,k^{3/2}e^{-\varepsilon k}\sin kr \\ &= \frac{3(\hbar c)^{1/2}}{8\pi^{3/2}}\frac{1}{r(r^2+\varepsilon^2)^{5/4}}\sin\left(\frac{5}{2}\arctan\frac{r}{\varepsilon}\right)\,. \end{aligned} \qquad (7)$$

When $\varepsilon \to 0$, or more generally whenever $\varepsilon \ll r$, this reduces to

$$G(\mathbf{r}) = \pm \frac{3(\hbar c)^{1/2}}{8\sqrt{2}\pi^{3/2}} r^{-7/2} \,. \tag{8}$$

For the derivations in the next sections it is useful to notice that when in the spherically symmetric case a 3D Fourier integral is transformed into corresponding 1D one, the factor $(kr)^{-1} \sin kr$ appears in the integrand. This factor represents spherically symmetric standing wave and the integral is nothing but a superposition over such waves of different frequencies.

One can conclude from Eqs. (5) and (8) that—citation from (Mandel & Wolf, 1995) follows—"even when the position wave function $\mathbf{\Phi}(\mathbf{r}, t)$ is strongly concentrated near the origin, the energy wave function is spread out over space asymptotically like $r^{-7/2}$. Alternatively, we may say that even when the probability distribution of the photon is strongly localized near the origin, the energy distribution extends over large distances and falls off as r^{-7}." As the EM field operators contain the same $\mathbf{k}$-dependent factor $\sqrt{\hbar\omega}$ and are proportional to $\mathbf{\Phi}_E(\mathbf{r}, t)$, the corresponding measurable quantities, incl. the probability of photodetection, bear the same non-local relation to the photon probability distribution $|\mathbf{\Phi}(\mathbf{r}, t)|^2$. Again, one can conclude that—citation from (Mandel & Wolf, 1995) continues – "for a photon which is strongly localized close to the origin, there is a non-vanishing probability falling off as r^{-7} that it will be detected by a photoelectric detector at a distance r."

3. Exponential localization

In this section we see that the statements cited in the previous paragraph are incorrect.

Due to the orthogonality of the polarization vectors $\varepsilon^*_{\mathbf{k},s} \cdot \varepsilon_{\mathbf{k},s'} = \delta_{ss'} (s, s' = 1, 2)$ the total energy of a one-photon state Eq. (4) breaks into two non-interfering contributions from both polarization states. Therefore one may consider only one polarization at a time (or assume $\phi(\mathbf{k}, s = 2) \equiv 0$). Thus, the problem of the best localization of a photon reduces to the question: what is the fastest possible falloff with the distance $\mathbf{r}$ of the modulus of vector function $\mathbf{\Phi}_E(\mathbf{r}, t)$?

The most consistent treatment of the photon (energy) wave function is based on the Riemann-Silberstein (RS) complex vector which is given (in the SI system) by the following linear combination of the EM field vectors (Bialynicki-Birula, 1996; 1998)

$$\mathbf{F}(\mathbf{r}, t) = \frac{\mathbf{D}(\mathbf{r}, t)}{\sqrt{2\varepsilon_0}} + i \frac{\mathbf{B}(\mathbf{r}, t)}{\sqrt{2\mu_0}} .$$

Upon quantization of the electromagnetic field, the RS vector becomes the field operator $\hat{\mathbf{F}}(\mathbf{r}, t)$. It can be most conveniently decomposed into circularly polarized plane-wave modes with polarization vectors $\mathbf{e}_\pm(\mathbf{k})$ for left-handed and right-handed photons, which are related as $\mathbf{e}_-(\mathbf{k}) = (\mathbf{e}_+(\mathbf{k}))^*$. Using the RS operator with such decomposition for expressing the energy density of a one-photon state, the energy breaks into two independent contributions from both polarizations already before integration over space, i. e., the energy density turns out to be given as

$$|\mathbf{\Phi}_E(\mathbf{r}, t)|^2 = |\mathbf{F}_+(\mathbf{r}, t)|^2 + |\mathbf{F}_-(\mathbf{r}, t)|^2 \,. \tag{9}$$

where

$$\mathbf{F}_\pm(\mathbf{r}, t) = \int d^3k \, d(k) \mathbf{e}_\pm(\mathbf{k}) f_\pm(\mathbf{k}) e^{i(\mathbf{kr} - \omega t)} \,. \tag{10}$$

Here $d(k)$ is the frequency dependent normalization factor $d(k) = (2\pi)^{-3/2}\sqrt{\hbar\omega} = \sqrt{\hbar c/(2\pi)^3}k^{1/2}$ and the weight functions $\phi(\mathbf{k}, s = 1, 2)$ have been expressed for the circularly polarized one-photon states as $(2\pi)^{-3/2} f_{\pm}(\mathbf{k})$, cf. Eq. (3).

Again, since both polarization states in Eq. (9) contribute independently (incoherently) to the total energy density, one may consider only one polarization at a time. Thus, the problem of the best localization of a photon reduces to the question: what is the fastest possible falloff with the distance $r = |\mathbf{r}|$ of the modulus of vector functions $\mathbf{F}_{\pm}(\mathbf{r}, t)$?

Since we are interested in a uniform localization in all directions (in 3D space, later in a 2D plane), the expansion in Eq. (10) has to contain plane waves with many very different directions of vectors $\mathbf{k}$. As the constituents of the expansion are transversal plane waves obeying the condition $\mathbf{k} \times \mathbf{e}_{+}(\mathbf{k}) = -ik\mathbf{e}_{+}(\mathbf{k})$, the polarization vectors $\mathbf{e}_{\pm}(\mathbf{k})$ correspondingly possess various directions as well—that highly complicates study of falloff properties of $|\mathbf{F}_{+}(\mathbf{r}, t)|$.

Fortunately, such study can be conveniently carried out if we express, following (Bialynicki-Birula, 1996; 1998), the RS vector in terms of a "superpotential" $\mathbf{Z}(\mathbf{r}, t)$,

$$\mathbf{F}(\mathbf{r}, t) = \nabla \times \left[i\frac{\partial}{\partial ct}\mathbf{Z}(\mathbf{r}, t) + \nabla \times \mathbf{Z}(\mathbf{r}, t) \right] . \tag{11}$$

The complex vector field $\mathbf{Z}(\mathbf{r}, t)$ is a complexified version of the Hertz potential, which must obey the homogeneous wave equation and therefore has the following decomposition into plane waves

$$\mathbf{Z}(\mathbf{r}, t) = \int d^3k \left[\mathbf{h}_{+}(\mathbf{k})e^{i(\mathbf{kr}-\omega t)} + \mathbf{h}_{-}^{*}(\mathbf{k})e^{-i(\mathbf{kr}-\omega t)} \right] , \tag{12}$$

where $\mathbf{h}_{\pm}(\mathbf{k})$ now are *arbitrary* vector functions of $\mathbf{k}$, whose directions are not governed by any transversality condition. Like in the case of well-known problem of dipole radiation we may take unidirectional set of vectors $\mathbf{h}_{\pm}(\mathbf{k})$ which depend only on the modulus of $\mathbf{k}$. In such case we stand a good chance of finding model closed-form expressions for the integral in Eq. (12) with the help of tables of integrals and/or transforms, as we see below Thereafter one can derive $\mathbf{F}_{+}(\mathbf{r}, t)$ (or $\mathbf{F}_{-}(\mathbf{r}, t)$) from Eq. (11) and study falloff behavior of photon energy density $|\mathbf{F}_{+}(\mathbf{r}, t)|^2$ (or $|\mathbf{F}_{-}(\mathbf{r}, t)|^2$), which is also proportional to the photon detection probability.

As we are interested in one polarization state we will deal with the first (positive frequency) term in Eq. (12) only and let us choose $\mathbf{h}_{+}(\mathbf{k})$ in the form (Bialynicki-Birula, 1998)

$$\mathbf{h}_{+}(\mathbf{k}) \equiv \mathbf{h}(\mathbf{k}) = \mathbf{m}l^3\frac{h(\sigma)}{\sigma} , \tag{13}$$

where $\mathbf{m}$ is a constant vector that includes the normalization factor and $\sigma = kl$ is wavenumber measured in units of a characteristic length l that will play the role of photon wave function falloff parameter, i. e., l controls the volume of spherically symmetric localization. Since the wave vector dependence of $\mathbf{h}(\mathbf{k})$ in Eq. (13) is isotropic, the 3D integral in Eq. (12) – like in derivation Eq. (7)—reduces into 1D integral in spherical coordinates yielding

$$\mathbf{Z}(\mathbf{r}, t) = 4\pi\mathbf{m}l^2 \int_0^\infty dk\, h(lk)\frac{\sin kr}{r}e^{-ikct} \tag{14}$$

$$= 2\pi i\, \mathbf{m}\frac{l}{r} \left[g(\frac{ct+r}{l}) - g(\frac{ct-r}{l}) \right] , \tag{15}$$

where the function g is given by the Fourier transform of $h(\sigma)$ over positive frequencies only

$$g(\tau) = \int_0^\infty d\sigma\, h(\sigma) e^{-i\sigma\tau} \,. \tag{16}$$

If we split the function under the transform into two factors $h(\sigma) = \bar{h}(\sigma)\exp(-\sigma)$, we notice that the Fourier integral can be expressed as the Laplace transform at a point in the complex plane, whose real coordinate has value 1

$$g(\tau) = \mathcal{L}\left\{\bar{h}(\sigma); \sigma, 1 + i\tau\right\} \,. \tag{17}$$

Eq. (17) opens possibility to carry out search of such functions $\bar{h}(\sigma)$ in rich tables of the Laplace transform, which correspond to strong falloff of $g(\tau)$. Indeed, for all spectra of the form

$$\bar{h}_{H_n}(\sigma) = 2^{-1}\sigma^{-\frac{n+1}{2}} \exp\left(-\frac{1}{\sigma}\right) H_n\left(\sqrt{\frac{1}{\sigma}}\right) \,, \tag{18}$$

which contains only half-integer negative powers of the wavenumber irrespective of the order n of the Hermite polynom H_n and due to the exponential factor approaches zero very rapidly as the wavenumber approaches zero, there is a closed-form Laplace transform in (Bateman & Erdelyi, 1954) Sect. 4.11, Eq. (18) or Sect. 5.6, Eq. (8), which yields

$$g_{H_n}(\tau) = 2^{n-1}\sqrt{\pi}\,(1+i\tau)^{\frac{n-1}{2}} \exp\left(-2\sqrt{1+i\tau}\right) \,, \tag{19}$$

or, consequently, with the help of Eq. (15)

$$\mathbf{Z}(\mathbf{r},t) = 2^n \pi^{3/2} i\, \mathbf{m} \frac{l}{r} \begin{bmatrix} \left(1 + i\frac{ct+r}{l}\right)^{\frac{n-1}{2}} \exp\left(-2\sqrt{1 + i\frac{ct+r}{l}}\right) - \\ - \left(1 + i\frac{ct-r}{l}\right)^{\frac{n-1}{2}} \exp\left(-2\sqrt{1 + i\frac{ct-r}{l}}\right) \end{bmatrix} \tag{20}$$

A particular case with $n = 1$, when

$$\bar{h}_1(\sigma) = \sigma^{-\frac{3}{2}} \exp\left(-\frac{1}{\sigma}\right) \,, \tag{21}$$

$$\mathbf{Z}(\mathbf{r},t) = 2\pi^{3/2} i\, \mathbf{m} \frac{l}{r} \left[\exp\left(-2\sqrt{1 + i\frac{ct+r}{l}}\right) - \exp\left(-2\sqrt{1 + i\frac{ct-r}{l}}\right)\right] \tag{22}$$

was found in (Bialynicki-Birula, 1998) and from Eq. (22) $\mathbf{Z}(\mathbf{r}, t = 0)$ and $i\partial_{ct}\mathbf{Z}(\mathbf{r}, t = 0)$, which turn out to be real quantities, were calculated there as well.

The Hertz vector given by Eqs. (20), (22) describes a broadband single photon state having the form of a spherical shell converging for negative and diverging for positive values of time t, attaining the maximal localization at instant $t = 0$. The function $\mathbf{Z}(\mathbf{r}, t)$ and its time derivative $\partial_{ct}\mathbf{Z}(\mathbf{r}, t)$ fall off exponentially at large r as $\exp(-\sqrt{2r/l})$ (multiplied by some power of r), and this property will be shared by their space derivatives involved in Eq. (11). Hence, the photon energy density as defined by Eq. (9) will also exhibit an exponential falloff.

Browsing tables of integrals reveals that for obtaining the exponential falloff the spectra need not contain negative powers of the wavenumber in combinations prescribed by special polynomials like in Eq. (18). For example, the spectra containing only a single half-integer negative power

$$\bar{h}_n(\sigma) = \sigma^{-n-\frac{1}{2}} \exp\left(-\frac{q}{\sigma}\right) , \tag{23}$$

(were q is an optional dimensionless parameter) with the help of (Gradshteyn & Ryzhik, 2000) Eq. (3.472-5) gives

$$g_n(\tau) = (-1)^n \sqrt{\frac{\pi}{1+i\tau}} \frac{\partial^n}{\partial q^n} \exp\left[-2\sqrt{q\,(1+i\tau)}\right] .$$

We see that by putting $q = 1$ we reach exactly the same exponential factor as in Eq. (19).

As shown in the previous Section, the spectrum of the Landau-Peierls (LP) wave function Eq. (2) contains factor $k^{-1/2}$ as compared to the spectrum of photon wave function Eq. (3) or Eq. (10) due to the extra factor $\sqrt{\hbar\omega}$ or $d(k)$ in the expansion of the energy wave function. Hence, in order to obtain the photon position wave function from the Hertz potential with the model spectra Eqs. (19), (21), (23) by the same procedure, one has to use the spectra with an additional factor $(\hbar c)^{-1/2}\,\sigma^{-1/2}$, i.e., the spectra with integer powers of the wavenumber. Specifically, instead of Eq. (23) we must start (if omitting constants and taking $q = 1$) with

$$\bar{h}_n^{LP}(\sigma) = \sigma^{-n-1} \exp\left(-\frac{1}{\sigma}\right) .$$

Fortunately, for such spectrum a closed-form Laplace transform exists (Bateman & Erdelyi, 1954) Eq. (5.16-40) yielding

$$\begin{aligned} g_n^{LP}(\tau) &= \\ &= \mathcal{L}\left(\sigma^{-\eta-1} \exp\left(-\frac{1}{\sigma}\right); \sigma, 1+i\tau\right) = 2\,(1+i\tau)^{\eta/2} K_\eta(2\sqrt{1+i\tau})\, , \; \eta = n\, , \end{aligned} \tag{24}$$

where K_η is the modified Bessel function (or the Macdonald function) of order η. This result can be used also for obtaining $g_{H_n}^{LP}(\tau)$ corresponding to energy wave function spectrum Eq. (18) if to write out explicitly the Hermite polynomial. Parenthetically, as the last transform formula is valid for any, even complex value of η, then for the half-integer value $\eta = 1/2$ we recover Eq. (22), since $K_{\pm\frac{1}{2}}(z) = \sqrt{\pi/2z}e^{-z}$. Moreover, for any, incl. integer order, the last equality holds *asymptotically*

$$\lim_{|1+i\tau|\to\infty} \mathcal{L}\left(\sigma^{-\eta-1} \exp\left(-\frac{1}{\sigma}\right); \sigma, 1+i\tau\right) = \sqrt{\pi}\,(1+i\tau)^{\eta/2-1/4} e^{-2\sqrt{1+i\tau}} .$$

For the particular case $\eta = 1$ this result coincides with (Bialynicki-Birula, 1998) Eq. (34) .

Hence, we have reached a general result: if the wavenumber spectrum of decomposition of the Hertz vector $\mathbf{Z}(\mathbf{r}, t)$ into plane waves has the form

$$h(k) \propto \left(\frac{1}{kl}\right)^{\alpha} \exp\left(-kl - \frac{1}{kl}\right)$$

then, irrespective of the value of power α, at large distances the falloff of $\mathbf{Z}(\mathbf{r}, t)$ as well as of wave functions $\mathbf{F}(\mathbf{r}, t)$, $\mathbf{\Phi}(\mathbf{r}, t)$ and corresponding photon energy and probability densities is dominantly governed by exponential factor, the exponent being proportional to square root of the distance.

Follows inescapable conclusion—for the first time made by (Bialynicki-Birula, 1998) on the basis of one particular spectrum—that the textbook statements, cited in the end of previous Section, are incorrect. Both functions—the position wave function and the energy wave function—may be strongly (= exponentially) concentrated near the origin. The incorrect statements seemingly stem from an idea that asymptotic behavior of convolution of two functions is governed by the one with slower falloff. Apparently such understanding is superficial and need not to be correct if none of the functions has finite support.

4. Limits of uniform localization in all directions

Having shown possibility of asymptotic falloff as $\sim \exp(-A\sqrt{r})$, where A is a constant, naturally the question arises whether a stronger localization is allowed, e. g., according to the exponent with a higher power of the radius r^γ, $1/2 < \gamma \geq 1$. The answer comes from inspection of Eq. (16) with the help of the Paley-Wiener theorem—or criterion, see (Bialynicki-Birula, 1998). According to this theorem, the Fourier transform $g(\tau)$ of a square-integrable function $h(\sigma)$ that vanishes for all negative values of σ (i. e., for negative frequencies in our context) must obey the following integrability condition:

$$\int_{-\infty}^{\infty} d\tau \, \frac{|\log |g(\tau)||}{1+\tau^2} < \infty \,. \tag{25}$$

This condition does not allow for the exponential falloff with $\gamma \geq 1$ but anything arbitrarily weaker than that is allowed. For example, $g(\tau) \sim \exp(-A\tau^{n/(n+1)})$ and even $g(\tau) \sim \exp(-A\tau/\log\tau)$, etc, i. e., almost linear exponential functions are allowed (Bialynicki-Birula, 1998).

Let us take a closer look at how the Paley-Wiener criterion restricts localization of a function whose spectrum contains only positive frequencies, i. e., constitutes an analytic signal.

An 1D right-moving wave function with white spectrum is ultimately (i. e., delta-) localized forever: $\Psi(x,t) \sim \delta(x-ct)$. If we cut off the negative frequency half of the spectrum, we get

$$\Psi_+(x,t) \sim \delta_-(x-ct)\,,$$
$$\delta_-(y) \equiv \frac{1}{2\pi}\int_0^\infty dk e^{iky} = \frac{1}{2}\left[\delta(x-ct) + \frac{i}{\pi} P \frac{A}{x-ct}\right],$$

where the principal value (P) term corrupts the delta-localization. Since $|\delta_-(y)| = 1/2\pi\,|y|$, the falloff is slow: reciprocally proportional to the distance from the wave peak. Parenthetically, two counterpropagating δ_--pulses colliding at the origin do not constitute an analytical wave (but still an analytic signal in respect of time) because the wavenumber takes now values of both signs. The imaginary part of such standing-wave-type wave function vanishes at the instant $t = 0$, i. e., it is delta-localized at that instant.

This simple example of ultimate localization in 1D helps to study the case of uniform localization in 3D space. Let us take the spectrum $h(lk)$ in the integral Eq. (14), which is nothing but a superposition of standing spherical waves, in the form $h(lk) =$

$4\pi^{-1}\sin^2(lk/2)/k$. Then for the instant $t = 0$ with the help of Eq. (3.828-3) of Ref. (Gradshteyn & Ryzhik, 2000) we get

$$\begin{aligned} \int_0^\infty dk\, 4\pi^{-1}k^{-1}\sin^2(lk/2)\sin kr &= 1, \qquad r < l, \\ &= 1/2, \qquad r = l, \\ &= 0, \qquad r > l\,. \end{aligned} \tag{26}$$

In other words, in the case of such spectrum $\mathbf{Z}(\mathbf{r}, t = 0)$ is confined into spherical cell of radius l. The support of $\mathbf{Z}(\mathbf{r}, t = 0)$ is finite and in this sense the localization is the strongest. There is no restrictions by the Paley-Wiener theorem here, because strictly at the instant $t = 0$ of maximal localization the integral is nothing but the sine transform for which the theorem does not apply. Indeed, the sine transform tables give examples of the resultant functions with arbitrarily abrupt falloff. However, it does not mean as if the photon localization restriction was lifted at the instant $t = 0$. The explanation is that according to Eq. (11) the energy wave function involves also the time derivative of $\mathbf{Z}(\mathbf{r}, t)$ at $t = 0$, but the sine transforms of two functions $h(lk)$ and $h(lk)k$ (the additional frequency factor k enters the integrand as the Fourier image of $\partial/\partial ct$) cannot simultaneously possess arbitrarily abrupt falloffs.

As soon as $t \neq 0$ Eq. (26) is replaced by the Fourier transform over positive frequencies, which can be evaluated through the Laplace transform $\mathcal{L}$ as follows

$$\begin{aligned} \mathbf{Z}(\mathbf{r}, t) &= 4\pi\mathbf{m}lr^{-1}I(r, t)\,, \\ I(r, t) &= 4\pi^{-1}\int_0^\infty dk\, k^{-1}\sin^2(lk/2)\sin kr\, e^{-ikct} \qquad (27) \\ &= 4\pi^{-1}\lim_{\varepsilon\to 0}\mathcal{L}\left\{k^{-1}\sin^2(lk/2)\sin kr; k, \varepsilon + ict\right\} \\ &= \pi^{-1}\lim_{\varepsilon\to 0}\left[\arctan\left(\frac{l-r}{\varepsilon + ict}\right) - \arctan\left(\frac{l+r}{\varepsilon + ict}\right) + 2\arctan\left(\frac{r}{\varepsilon + ict}\right)\right], \qquad (28) \end{aligned}$$

where ε is the real part of the Laplace transform variable. As one can see in Fig. 1, falloff of $I(r, t \neq 0)$ obeys the Paley-Wiener restriction indeed—asymptotically it is slower than exponential decay with linear exponent. Naturally, $I(r, t = 0)$ returns the behavior of Eq. (26), i. e., the strict confinement into spherical cell of radius l. Sharp peaks and long tails of the modulus of $I(r, t \neq 0)$ originate from the imaginary part of $I(r, t)$ which is—as it is known for an analytic signal—related to the real part through the Hilbert transform. The latter resembles operation of taking derivative but is non-local, thus explaining appearance of the peaks and tails.

Let us take a closer look at the real part of $I(r, t)$, because it deserves interest not only as the main contribution to $|I(r, t)|$. Common classical Hertz potential is a real quantity and contains negative frequencies $\omega = -kc$ as well. Since k as wavenumber in spherical coordinates is by definition positive everywhere in the integrand of Eq. (27) except in the exponent, where it stands for the frequency $\omega = \pm kc$, integration over negative frequencies as well is equivalent to adding to the integral its complex conjugate $I^*(r, t)$. Hence, expressions for classical fields would be governed by the real part of Eq. (28), which is free from the Paley-Wiener restrictions. As one can see in Fig. 2, it constitutes a spherical bipolar pulse of rectangular profile, which collapses—with negative half-cycle ahead– to the origin at negative times and expands – with negative half-cycle behind—from the origin at positive times. At all times it preserves

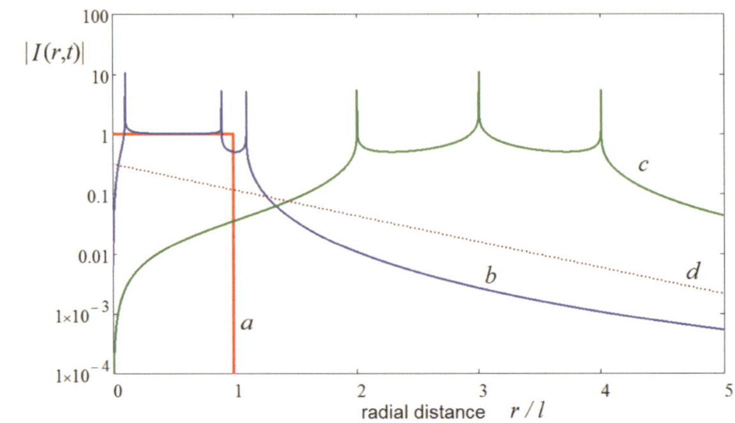

Fig. 1. (color online). Radial dependence of Eq.(28) at various time instants in units l/c: *a* (red line), $t = 0$; *b* (blue), $t = 0.1$; *c*(green), $t = 3$. For comparison falloff of $\pi^{-1}\exp(-r/l)$ is shown (dotted line *d*).

strict confinement. Between the stages of collapse and expansion the profile undergoes a transformation which is characteristic for all few- and sub-cycle focusing pulses and is caused by the Gouy phase shift (Saari, 2001). Having in mind that sections along horizontal lines in Fig. 2 give radial dependences at fixed instants like in Fig. 1, one can verify that the sharp peaks in the imaginary part correspond indeed to the abrupt steps in the real part.

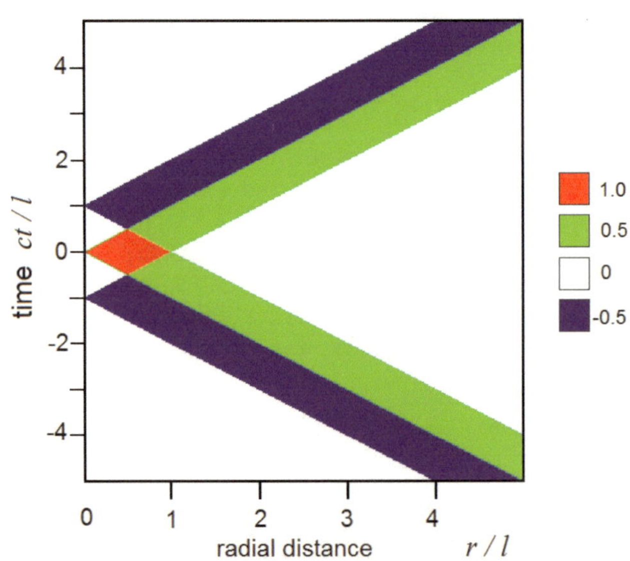

Fig. 2. (color online). Four value levels of the real part of $I(r,t)$ depicted in 2D plane of its arguments. In white areas the function Re$I(r,t)$ equals strictly to zero, i. e., its support has finite volume.

5. Localization in two dimensions

What if we allow nonuniform localization? Is a two-dimensional localization of the one-photon state also restricted with the Paley-Wiener criterion? As shown in (Saari et al., 2005) the answer to the latter question is NO, if we construct the wave functions from certain so-called localized waves (Besieris et al., 1998; Bialynicki-Birula & Bialynicka-Birula, 2006; Hernández-Figueroa et al., 2008; Recami et al., 2003; Saari & Reivelt, 2004; Salo et al., 2000), which are recently discovered solutions to the linear wave equation. This Section reproduces examples from (Saari et al., 2005).

As the first example leading to a stronger localization that one might expect from the Paley-Wiener theorem, we will consider the photon field which is a superposition of cylindrical solutions of the wave equation. Let us again use the Hertz potential approach, this time putting $\mathbf{Z}(\mathbf{r},\tau) = \mathbf{m}\Psi(\mathbf{r},\tau)$, where $\tau \equiv ct$, $\mathbf{m}$ is again a constant vector that includes the proper normalization factor and is directed along the axis z (any other orientation gives similar results), and $\Psi(\mathbf{r},\tau)$ is a wavepacket of Bessel functions J_0 with the exponential spectrum and a specific dispersion law for the axial wavenumber $k_z(\omega) = const = k_0$

$$\Psi(\rho,z,\tau) = \Delta \int_{|k_0|}^{\infty} dk\, J_0\left(k_\rho \rho\right) e^{-k\Delta} e^{-i(k\tau - k_0 z)}\,, \tag{29}$$

where the radial coordinate ρ has been introduced and $k_\rho = \left(k^2 - k_0^2\right)^{1/2}$ is the lateral component of the wave vector of the monochromatic plane-wave constituents represented with the weight function $e^{-k\Delta}$ whose width is Δ^{-1} (spectral width of the packet). The integral can be taken with the help of a Laplace transform table and we obtain

$$\mathbf{Z}(\rho,z,\tau) = \mathbf{m}\Delta \frac{\exp\left(-|k_0|\sqrt{\rho^2 + (\Delta + i\tau)^2}\right)}{\sqrt{\rho^2 + (\Delta + i\tau)^2}} e^{ik_0 z}\,. \tag{30}$$

Eq. (30) describes a simple cylindrical pulse modulated harmonically in the axial direction and radially converging (when $\tau < 0$) to the axis and thereafter (when $\tau > 0$) expanding from it, the intensity distribution resembling an infinitely long tube coaxial with the z axis and with a time-dependent diameter (see Fig. 6. in (Saari & Reivelt, 2004)). It follows from Eqs. (30) and (11) that

$$|\mathbf{Z}(\rho \to \infty, z, \tau = 0)| \sim \rho^{-1} \exp(-\rho/l)\,, \tag{31}$$

$$|\mathbf{F}(\rho \to \infty, z, \tau = 0)|^2 \sim \left[\rho^{-2} + O(\rho^{-3})\right] \exp(-2\rho/l)\,, \tag{32}$$

where $l \equiv |k_0|^{-1}$ is the characteristic length (or length unit). Thus, while the photon is completely delocalized in the axial direction, its energy density falloff in the lateral directions is exponential with the linear exponent at all times the conditions $\tau \ll \rho \gg \Delta$ are fulfilled, see Fig. 3. The time derivative as well as the spatial derivatives contain the same exponential factor, ensuring the exponential falloff of the the Riemann-Silberstein vector in Eq. (32).

Hence, a one-photon field given by Eq. (30) serves as the first and simplest example where the localization in two transversal dimensions is governed by different rules than uniform localization in three dimensions.

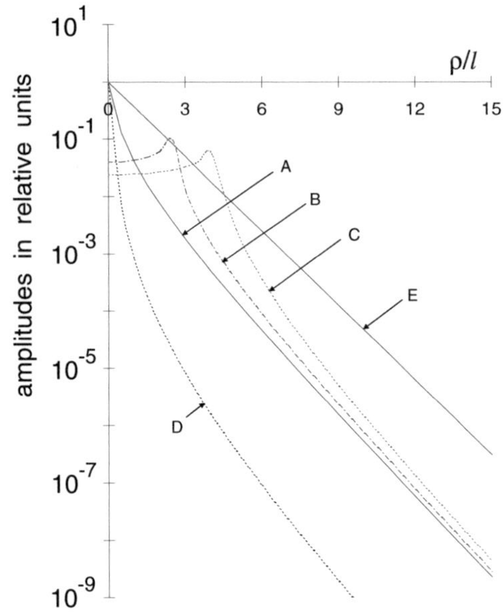

Fig. 3. Curves of the radial dependence in a decimal logarithmic scale. Curve A is for $|\mathbf{Z}(\rho,0,\tau=0)|$; B, $|\mathbf{Z}(\rho,0,\tau=2.5l)|$; C is the same as B but with Ψ taken from Eq. (33); D, $\left|\frac{\partial}{\partial\tau}\mathbf{Z}(\rho,0,\tau=0)\right|$; E is a reference curve $\exp(-\rho/l)$. The curves A, B, and C have been normalized so that $|\mathbf{Z}(0,0,0)|=1$. The values of the remaining free parameters are $\Delta=0.1l$ and $\beta=0.8$.

The next example is readily available via the Lorentz transformation of the wave function given by Eq. (29) along the axis z, which gives another possible solution of the scalar wave equation. The result is a new independent solution but it can also be considered as the wave given by Eqs. (29) and (30), which is observed in another inertial reference frame (Saari & Reivelt, 2004):

$$\Psi(\rho,z,\tau)=\Delta\frac{\exp\left(-|k_0|\sqrt{\rho^2+(\Delta-i\gamma(\beta z-\tau))^2}\right)}{\sqrt{\rho^2+(\Delta-i\gamma(\beta z-\tau))^2}} \times\exp\left(i\gamma k_0(z-\beta\tau)\right), \tag{33}$$

where the relativistic factors $\gamma\equiv(1-\beta^2)^{-1/2}$ and $\beta\equiv v/c<1$ have been introduced, v being a free parameter—the relative speed between the frames. In the waist region (see Fig. 4) this wave function has the same radial falloff as was given by Eq. (31), see curve "C" in Fig. 1, while the axial localization follows a power law. The strongly localized waist and the whole amplitude distribution move rigidly and without any spread along the axis z with a superluminal speed c/β. Such wave with intriguing properties, named the focused X wave (FXW) (Besieris et al., 1998; Saari & Reivelt, 2004), belongs to the so-called superluminal propagation-invariant localized waves. Although the FXW is not experimentally generated yet, a set-up based on a cylindrical diffraction grating has been proposed and its properties analyzed (Valtna et al., 2007). It should be noted here that there is nothing unphysical in the superluminality of the localized waves since a superluminal group velocity does not mean as if energy or information could be transmitted faster than c. This is an experimentally verified fact for the so-called Bessel-X pulse which is another representative of the family of superluminal waves (Alexeev et al., 2002; Bowlan et al., 2009; Saari et al., 2010; Saari & Reivelt, 1997). For a detailed analysis of the startling superluminality see (Saari, 2004; Saari et al., 2010) and references therein.

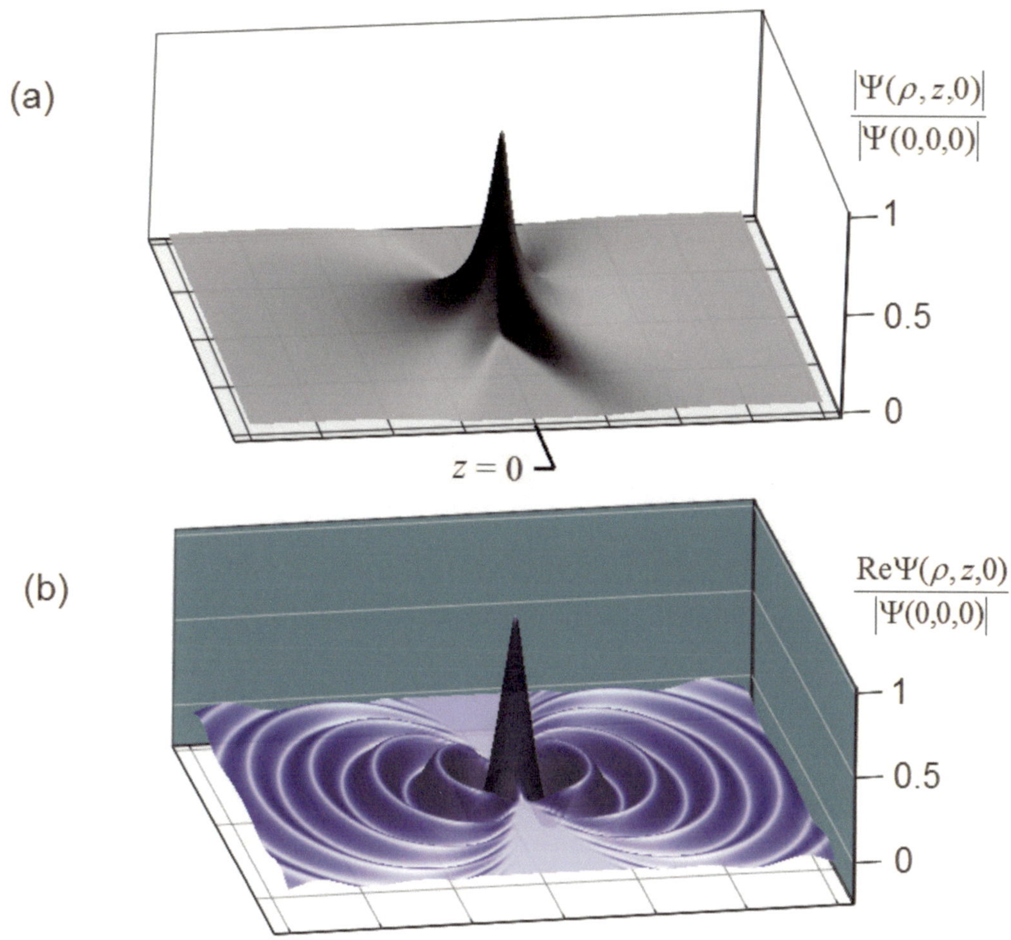

Fig. 4. (color online). The superluminal FXW given by Eq. (33). Shown are the dependences (a) of the modulus and (b) of the real part of the wavefunction on the longitudinal (z, increasing to the right) and a transverse (say, x) coordinates. The distance between the grid lines on the basal plane (x, z) is 22λ, where $\lambda = 2\pi|k_0|^{-1}$, k_0 being negative. The values of the remaining free parameters are $\Delta = 30\lambda$ and $\beta = 0.995$ or $\gamma = 10$.

Hence, in its waist (cross-sectional) plane a one-photon field given by the FXW possesses the same strong localization at any time as the previously considered cylindrical field does in any transversal plane at the instant $t = 0$.

By making use of the historically first representative of localized waves – the so-called focus wave mode (FWM)—see (Brittingham, 1983; Sezginer, 1985) and also reviews (Besieris et al., 1998; Saari & Reivelt, 2004) and references therein – one readily obtains an example of the field that exhibits even much stronger than simple exponential localization. The FWM is given by the scalar function

$$\Psi(\rho, z, \tau) = a \frac{\exp\left[-\frac{\rho^2}{2l(a-i(z-\tau))}\right]}{a - i(z-\tau)} \exp\left[-\frac{i(z+\tau)}{2l}\right], \tag{34}$$

where again l is a wavelength-type characteristic length and the constant a controls the axial localization length. This wave function is depicted in Fig. 5. Since the FXW in the limit $\beta \to 1$ becomes a FWM (Besieris et al., 1998; Saari & Reivelt, 2004), Fig. 5 qualitatively resembles Fig. 4(b) (a 3D animated color plot of FWM is avalable in open-access on-line paper (Sheppard & Saari, 2008)).

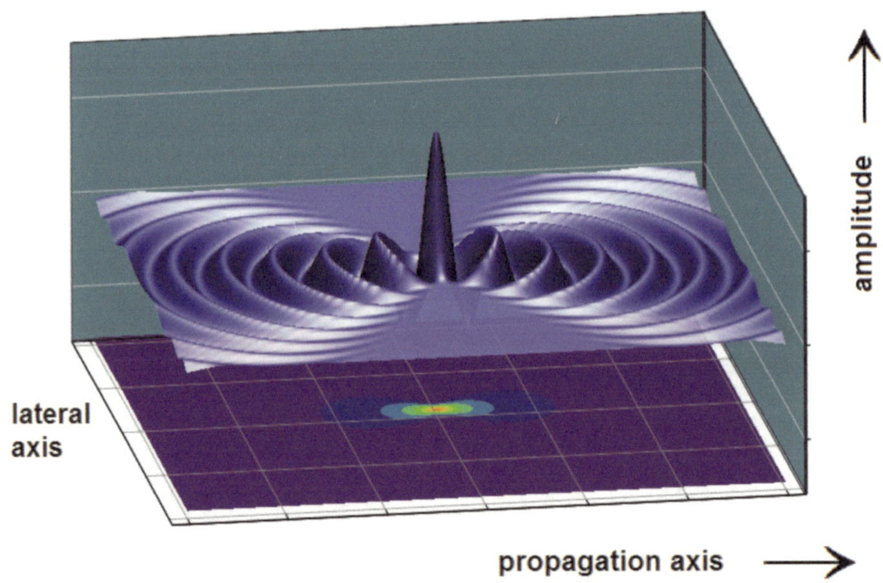

Fig. 5. (color online). The luminal FWM given by Eq. (34). Shown are a surface plot of the real part of the wave function and (in the basement plane) a contour plot of its modulus. For details and animation of the time dependece see (Sheppard & Saari, 2008), http://www.opticsinfobase.org/oe/viewmedia.cfm?uri=oe-16-1-150&seq=3 .

Multiplying Eq. (34) by $\mathbf{m}$ to build the vector $\mathbf{Z}(\rho, z, \tau)$ and inserting the latter into Eq. (11) we obtain that in this example the photon localization in the waist plane is quadratically exponential (Gaussian falloff):

$$|\mathbf{Z}(\rho \to \infty, z = \tau)| \sim \exp(-\rho^2/2la), \tag{35}$$

$$|\mathbf{F}(\rho \to \infty, z = \tau)|^2 \sim \rho^6 \exp(-\rho^2/la) \,. \tag{36}$$

In Eq. (36) only the highest-power term with respect to ρ is shown.

6. Discussion

To start discussing our results that seem to be at variance with the Paley-Wiener restriction, let us ask first whether the wave functions considered are something extraordinary. The answer is: yes, they are indeed, since browsing various integral transform tables reveals rather few examples where both the real and imaginary part of a wave function and of its time derivative have simultaneously an exponential or stronger localization in conjunction with other requisite properties. Fortunately, the list of proper wave functions with an extraordinary strong localization is growing—in addition to an optically feasible version (Reivelt & Saari, 2002; 2004) of the FWM various new interesting solutions with Gaussian falloff can be derived (Kiselev, 2007). Yet, it could be argued that the well-known Gaussian beam pulse has the same quadratically exponential radial profile in the waist region. However, resorting to the family of the Gaussian beams (the Gauss-Laguerre and Gauss-Hermite beams, etc.) is irrelevant here. The reason is that all these beams are solutions of the wave equation only in the paraxial approximation not valid in the case of any significant localization of wide-band (pulsed) superpositions of the beams and at large values of the radial distance. As a matter of fact, e.g., an exact solution corresponding to a lowest-order (axisymmetric) Gaussian beam has a weak power-law radial falloff in the waist region (Saari, 2001; Sheppard & Saghafi, 1999).

The next possible objection to the physical significance of the results obtained might arise from the infinite total energy (Besieris et al., 1998; Hernández-Figueroa et al., 2008) of the waves given by Eqs. (30), (33), and (34). However, at any spatial location the wave function is square integrable with respect to time, thus the condition of the Paley-Wiener theorem has been satisfied. Moreover, physically feasible finite-energy (i. e., finite-aperture) versions of localized waves generally exhibit even better localization properties, although not persistently. A finite-energy version of the FXW, called the modified focused X wave (MFXW) (Besieris et al., 1998; Valtna et al., 2006), has the same exponential factor as in Eq. (33), which is multiplied by a fraction that allows to force the axial localization to follow an arbitrarily strong power-law. The latter circumstance indicates that the strong lateral localization of the fields considered does not appear somehow at the expense of their axial localization.

As a matter of fact, energy-normalization of a wave function depends on how many photons it describes. It is easy to see that derivations and results presented here hold for any number state with $N \geq 1$ and also for incoherent mixtures of such states (which is important for experimental studies). Here it is not of interest to consider coherent states since generally for states of electromagnetic field that have classical counterparts one can escape—already in the case of uniform spherical localization—the constraints imposed by the Paley-Wiener theorem (Bialynicki-Birula, 1998).

The final crucial question is: are our results in contradiction with those of (Bialynicki-Birula, 1998) reproduced in Section 3? The answer is *no*, since in the case of the cylindrical waves the radial distance and temporal frequency are not directly Fourier-conjugated variables. In order to clarify this point, let us first take a closer look at Eqs. (14) and (15). The sine in Eq. (14) results from the imploding and exploding spherical wave constituents of the standing wave, like an odd one-dimensional standing wave arises from counterpropagating waves. We saw in Section 4 that although strictly at the instant $\tau = 0$ the function $\mathbf{Z}(\mathbf{r}, t)$ can possess arbitrarily abrupt falloff, simultaneously its time derivative and hence the wave function $\mathbf{F}(\mathbf{r}, t)$ cannot. In contrast, the time derivative of the wave function given by Eq. (30) or Eq. (29) has the same strong exponential falloff as the function itself, which persists for some (not too long) time, see Fig. 3. By comparing Eqs. (29) and (14) we notice that while in Eq. (14) the argument of the sine function is the product of the distance with the Fourier variable, in Eq. (29) the argument of the Bessel function is the product of the radial distance ρ with the radial wavenumber k_ρ the latter depending on the Fourier variable through the square-root expression with the constant parameter k_0—the lower limit of the integration. As it follows also from Eqs. (31) and (32) the condition $k_0 \neq 0$ is crucial for obtaining the exponential falloff. Hence, in the case of the cylindrical waves considered by us, the apparent violation of the rules set by the Paley-Wiener theorem results from the specific complicated relation between the radial distance and the Fourier variable.

7. Conclusion

The problem of photon localization is of rather fundamental nature in quantum electrodynamics. Despite of almost 80-year history of the problem – and the related problem of the photon wave function—the interest in the revision of it has quickened in the recent years. One of the stimulus for that might be developments in modern optics, particularly in femtosecond and quantum optics, thanks to which the somewhat academic problem is transforming into a practical one. Indeed, e. g., availability and applications of single- and sub-cycle photon pulses will force a revision of traditional notions in optics based on the narrow-band approximation. In particular, phrases like "localization cannot be better than wavelength" are loosing sense in the case of such pulses.

Ultrawideband by definition are the so-called localized waves—an emerging new field in wave acoustics and physical optics. We have shown that an interdisciplinary "technology transfer"—application of methods and solutions found in the field of localized waves—is productive for the study of photon localization.

8. Acknowledgements

The author is thankful to Iwo Bialynicki-Birula for stimulating hints, numerous discussions, and remarks. The research was supported by the Estonian Science Foundation. Its publication was supported by the European Regional Development Fund.

9. References

Akhiezer, A. & Berestetskii, V. (1965). *Quantum Electrodynamics*, Interscience.

Alexeev, I., Kim, K. & Milchberg, H. (2002). Measurement of the superluminal group velocity of an ultrashort Bessel beam pulse, *Phys. Rev. Lett.* 88(7): 073901.

Bateman, H. & Erdelyi, A. (1954). *Tables of integral transforms*, McGraw-Hill.

Belgiorno, F., Cacciatori, S. L., Clerici, M., Gorini, V., Ortenzi, G., Rizzi, L., Rubino, E., Sala, V. G. & Faccio, D. (2010). Hawking Radiation from Ultrashort Laser Pulse Filaments, *Phys. Rev. Lett.* 105(20): 203901.

Besieris, I., Abdel-Rahman, M., Shaarawi, A. & Chatzipetros, A. (1998). Two fundamental representations of localized pulse solutions to the scalar wave equation, *Progress In Electromagnetics Research* 19: 1–48.

Besieris, I. M., Shaarawi, A. M. & Ziolkowski, R. W. (1994). Nondispersive accelerating wave packets, *Am. J. Phys.* 62(6): 519–521.

Bialynicki-Birula, I. (1996). Photon wave function, *in* E. Wolf (ed.), *Progress in Optics, vol 36*, North-Holland, Amsterdam, pp. 1–38.

Bialynicki-Birula, I. (1998). Exponential localization of photons, *Phys. Rev. Lett.* 80(24): 5247–5250.

Bialynicki-Birula, I. & Bialynicka-Birula, Z. (2006). Beams of electromagnetic radiation carrying angular momentum: The Riemann-Silberstein vector and the classical-quantum correspondence, *Opt. Comm.* 264(2): 342–351.

Bialynicki-Birula, I. & Bialynicka-Birula, Z. (2009). Why photons cannot be sharply localized, *Phys. Rev. A* 79(3): 032112.

Bowlan, P., Valtna-Lukner, H., Lõhmus, M., Piksarv, P., Saari, P. & Trebino, R. (2009). Measuring the spatiotemporal field of ultrashort Bessel-X pulses, *Opt. Lett.* 34(15): 2276–2278.

Brittingham, J. (1983). Focus waves modes in homogeneous Maxwell equations - transverse electric mode, *Journal of Applied Physics* 54(3): 1179–1189.

Chan, K., Law, C. & Eberly, J. (2002). Localized single-photon wave functions in free space, *Phys. Rev. Lett.* 88(10): 100402.

Ciattoni, A. & Conti, C. (2007). Quantum electromagnetic X waves, *JOSA B* 24(9): 2195–2198.

Gradshteyn, I. & Ryzhik, I. (2000). *Tables of integrals, series, and products, 6th edition*, Academic Press.

Grunwald, R., Kebbel, V., Griebner, U., Neumann, U., Kummrow, A., Rini, M., Nibbering, E., Piche, M., Rousseau, G. & Fortin, M. (2003). Generation and characterization of spatially and temporally localized few-cycle optical wave packets, *Phys. Rev. A* 67(6): 063820.

Hawton, M. (1999). Photon wave functions in a localized coordinate space basis, *Phys. Rev. A* 59(5): 3223–3227.

Hawton, M. (2007). Photon wave mechanics and position eigenvectors, *Phys. Rev. A* 75(6): 062107.

Hernández-Figueroa, H. E., Zamboni-Rached, M. & Recami, E. (eds) (2008). *Localized waves*, Wiley.

Jáuregui, R. & Hacyan, S. (2005). Quantum-mechanical properties of bessel beams, *Phys. Rev. A* 71(3): 033411.

Keller, O. (2000). Space-time description of photon emission from an atom, *Phys. Rev. A* 62(2): 022111.

Keller, O. (2005). On the theory of spatial localization of photons, *Physics Reports* 411(1): 1–232.

Kiselev, A. P. (2007). Localized light waves: Paraxial and exact solutions of the wave equation (a review), *Optics and Spectroscopy* 102(4): 603–622.

Mandel, M. & Wolf, E. (1995). *Optical coherence and quantum optics*, Cambridge University Press.

Recami, E., Zamboni-Rached, M., Nobrega, K., Dartora, C. & Hernandez, H. (2003). On the localized superluminal solutions to the Maxwell equations, *IEEE Journal of Selected Topics in Quantum Electronics* 9(1): 59–73.

Reivelt, K. & Saari, P. (2002). Experimental demonstration of realizability of optical focus wave modes, *Phys. Rev. E* 66(5): 056611.

Reivelt, K. & Saari, P. (2004). Bessel-Gauss pulse as an appropriate mathematical model for optically realizable localized waves, *Opt. Lett.* 29(11): 1176–1178.

Saari, P. (2001). Evolution of subcycle pulses in nonparaxial Gaussian beams, *Opt. Express* 8(11): 590–598.

Saari, P. (2004). Localised waves in femtosecond optics, *in* Miller, A and Reid, DT and Finlayson, DM (ed.), *Ultrafast Photonics*, Institute of Physics Conference Series, pp. 317–340. 56th Scottish Universities Summer School in Physics on Ultrafast Photonics (SUSSP56)NATO, Univ St Andrews, St Andrews, Scotland, SEP 01-14, 2002.

Saari, P., Bowlan, P., Valtna-Lukner, H., Lõhmus, M., Piksarv, P. & Trebino, R. (2010). Basic diffraction phenomena in time domain, *Opt. Express* 18(11): 11083–11088.

Saari, P., Menert, M. & Valtna, H. (2005). Photon localization barrier can be overcome, *Opt. Comm.* 246(4-6): 445–450.

Saari, P. & Reivelt, K. (1997). Evidence of X-shaped propagation-invariant localized light waves, *Phys. Rev. Lett.* 79(21): 4135–4138.

Saari, P. & Reivelt, K. (2004). Generation and classification of localized waves by Lorentz transformations in Fourier space, *Phys. Rev. E* 69(3, Part 2): 036612.

Salo, J., Fagerholm, J., Friberg, A. & Salomaa, M. (2000). Unified description of nondiffracting X and Y waves, *Phys. Rev. E* 62(3, Part b): 4261–4275.

Sezginer, A. (1985). A general formulation of focus wave modes, *Journal of Applied Physics* 57(3): 678–683.

Sheppard, C. J. R. & Saari, P. (2008). Lommel pulses: An analytic form for localized waves of the focus wave mode type with bandlimited spectrum, *Opt. Express* 16(1): 150–160.

Sheppard, C. & Saghafi, S. (1999). Electromagnetic Gaussian beams beyond the paraxial approximation, *JOSA A* 16(6): 1381–1386.

Sõnajalg, H., Rätsep, M. & Saari, P. (1997). Demonstration of the Bessel-X pulse propagating with strong lateral and longitudinal localization in a dispersive medium, *Opt. Lett.* 22(5): 310–312.

Valtna, H., Reivelt, K. & Saari, P. (2006). Modifications of the focused X wave as suitable models of strongly localized waves for realization in the optical domain, *Journal of Optics A* 8(2): 118–122.

Valtna, H., Reivelt, K. & Saari, P. (2007). Methods for generating wideband localized waves of superluminal group velocity, *Opt. Comm.* 278(1): 1–7.

4

Fusion Frames and Dynamics of Open Quantum Systems

Andrzej Jamiołkowski
Institute of Physics, Nicholas Copernicus University, 87–100 Toruń Poland

1. Introduction

Heisenberg's uncertainty principle is one of the manifestations of quantum complementarity. In particular, it states that upon measuring both the momentum and the position of a particle, the product of uncertainties has a fundamental lower bound proportional to Planck's constatnt. Hence, one cannot measure position and momentum simultaneously with a prescribed accuracy. In general, the quantum complementarity principle does not permit to identify a quantum state from measurements on a single copy of the system unless some extra knowledge is available.

One of the consequences of fundamental assumptions of quantum mechanics is the fact that determination of an unknown state can be achieved by appropriate measurements only if we have at our disposal a set of identically prepared copies of the system in question. Moreover, to devise a successful approach to the above problem of state reconstruction one has to identify a collection of observables, so-called *quorum*, such that their expectation values provide the complete information about the system state.

The problems of state determination have gained new relevance in recent years, following the realization that quantum systems and their evolutions can perform practical tasks such as teleportation, secure communication or dense coding. It is important to realize that if we identify the quorum of observables, then we also have a possibility to determine expectation values of physical quantities (observables) for which *no* measuring apparatuses are available.

Quantum tomography is a procedure of reconstructing the properties of a quantum object on the basis of experimentally accessible data. This means that quantum tomography can be classified by the type of object to be reconstructed:

1. *state tomography* treats density operators, which describe states of quantum systems;
2. *process tomography* discusses linear trace-preserving completely positive maps;
3. *device tomography* treats quantum instruments, and so on.

In what follows, we briefly describe the theory of quantum state tomography (cf. e.g. (Nielsen & Chuang, 2000; Weigert, 2000)).

The aim of quantum state tomography is to identify the density operator characterizing the state of a quantum system under consideration. Let $\mathcal{H}$ and $\mathcal{S}(\mathcal{H})$ denote the Hilbert space corresponding to the system and the set of all density operators on $\mathcal{H}$, respectively. We assume that the dimension of $\mathcal{H}$ is finite, $\dim \mathcal{H} = N$. According to the famous Born rule, if an observable corresponding to a Hermitian operator Q with discrete spectrum is measured in a system whose state is given by the vector $|\psi\rangle$, then 1) the measured result will be one of the eigenvalues λ of Q, and 2) the probability of measuring a given eigenvalue λ_i will be $\langle\psi|P_i|\psi\rangle$, where P_i denotes the projection onto the eigenspace of Q corresponding to λ_i. These statements are based on the existence of the spectral resolution for any observable Q. However, if Q is given as a square matrix of order $N > 4$, then it is well known that the problem of calculation of eigenvectors and eigenvalues of Q over the field $\mathbb{C}$ of complex numbers is *not* solvable by radicals in the general case. Even more, it is not solvable by any finite procedure in the situation, where only arithmetic operations are allowed. This means that, in fact, for a given Q we are not able to find effectively the spectral decomposition $Q = \sum \lambda_i P_i$. Therefore, we will suppose that the information about the state $\rho \in \mathcal{S}(\mathcal{H})$ is extracted from the expectation values of some observables $Q_1, \ldots, Q_r$, i.e.,

$$q_i = \mathrm{Tr}(\rho Q_i), \tag{1}$$

where q_i are real numbers inferred from the measurement and Q_i are self-adjoint operators on $\mathcal{H}$. (We do not assume the knowledge of spectral decompositions for Q_i.)

The question, how to construct a quorum of meaningful observables for a given quantum state is quite fundamental. Usually, one can identify only a small number of observables $Q_1, \ldots, Q_r$, where $r \ll N^2$, with clear physical meaninig, and their expectation values are not enough for the determination of a quantum state. As a natural remedy for this situation we can ask about the results of the measurements of these observables (their mean values) at different time instants $t_1, \ldots, t_s$ during the time evolution of the system in question (Jamiołkowski, 1982; 1983).

Summing up, as the fundamental objects in modern quantum theory one considers the set of states

$$\mathcal{S}(\mathcal{H}) := \{\rho : \mathcal{H} \to \mathcal{H};\ \rho \geq 0,\ \mathrm{Tr}\,\rho = 1\}, \tag{2}$$

and the set of bounded hermitean (self-adjoint) operators

$$\mathcal{B}_* := \{Q : \mathcal{H} \to \mathcal{H};\ Q = Q^*\}. \tag{3}$$

Time evolutions of systems are governed by linear master equations of the form (in the so-called Schrödinger picture)

$$\frac{d\rho(t)}{dt} = \mathbb{K}\,\rho(t), \tag{4}$$

or in the dual form (in the so-called Heisenberg picture)

$$\frac{dQ(t)}{dt} = \mathbb{L}\,Q(t), \tag{5}$$

where superoperators $\mathbb{K}$ and $\mathbb{L}$ act on operators from the sets $\mathcal{S}(\mathcal{H})$ and $\mathcal{B}_*(\mathcal{H})$, respectively. They represent dual forms of the same physical idea. Both sets $\mathcal{S}(\mathcal{H})$ and $\mathcal{B}_*(\mathcal{H})$ can be considered as subsets of the vector space $\mathcal{B}(\mathcal{H})$ of all bounded linear operators on $\mathcal{H}$ and they can be treated as scenes on which problems of quantum mechanical systems should be discussed.

Since in this paper we will consider finite-dimensional Hilbert spaces, therefore in fact $\mathcal{B}(\mathcal{H})$ denotes the set of *all* linear operators on $\mathcal{H}$. If we introduce in $\mathcal{B}(\mathcal{H})$ the scalar product by the equality

$$\langle A, B\rangle := \mathrm{Tr}(A^* B), \tag{6}$$

then $\mathcal{B}(\mathcal{H})$ can be regarded as yet another inner product space, namely the so-called *Hilbert-Schmidt space*. It is not difficult to see that $\mathcal{B}_*(\mathcal{H})$ with scalar product defined by (6) is a *real* vector space and $\dim \mathcal{B}_*(\mathcal{H}) = N^2$.

If one does not intend to describe the full dynamics but instead to give a "snapshot" of its effect at a particular time instant t, then one introduces the idea of a *quantum channel* which mathematically is represented by a completely positive trace preserving (CPTP) map. A completely positive map (a superoperator) is a transformation on density operators defined by the expression

$$\tilde{\rho} = \Phi(\rho(0)) = \sum_i A_i^* \rho(0) A_i , \tag{7}$$

where $A_i \in \mathcal{B}(\mathcal{H})$ are called *Kraus operators* (Kraus, 1971) or *noise operators* of the map Φ. The trace preservation condition implies that

$$\sum_i A_i A_i^* = \mathbb{I} . \tag{8}$$

Let us observe that a unitary evolution is a spacial case of the CPTP transformation, where there is only one unitary Kraus operator.

According to one of fundamental postulates of quantum theory one assumes that measurements change the state of the system in a way radically different from unitary evolution. The process of making a *von Neumann measurement* is formally described by an expression of the form (7) with the Kraus operators being some commuting self-adjoint idempotent operators P_i with the property $\sum P_i = \mathbb{I}$. A more general concept of measurement was introduced in the 1970-s by Davies and Lewis. This concept is formally expresses as a positive operator-valued measure (POVM) which is defined as a set of positive semidefinite operators $\{M_k\}$ satisfying $\sum M_k = \mathbb{I}$ and, obviously, every such M_k can be expressed in the form $M_k = F_k F_k^*$ (cf. e.g. Nielsen & Chuang, 2000). The operators M_k need not commute, and the result of a particular measurement depends, in general, on the order in which the measurements of M_k are performed.

The idea of stroboscopic tomography for open quantum systems appeared for the first time in the beginning of 1980's (although expressed in different terms (Jamiołkowski, 1982; 1983; 1986)). The main motivation came from quantum optics and the theory of lasers. In particular, using the concept of observability, in (Jamiołkowski, 1983) and (Jamiołkowski, 1986) the question of the minimal number of observables $Q_1, \ldots, Q_\eta$ for which the quantum systems can be $(Q_1, \ldots, Q_\eta)$-reconstructible was discussed.

On the other hand, theory of *frames*, which are collections of vectors that provide robust and usually non-unique representations of vectors, has been the subject of research in last decades and has been applied in these disciplines where redundancy played a vital and useful role.

However, in some applications it is natural to model and describe considered systems by collections of families of subspaces, and to split a large (global) frame system into a set of much smaller frame systems in these subspaces. This has led to the development of a suitable theory based on *fusion frames* (families of subspaces), which provides the framework to model these more complex applications (Casazza & Kutyniok, 2004; Casazza et al., 2008). In particular, a sequence of the so-called *k-order Krylov subspaces* which appear naturally in stroboscopic tomography (Jamiołkowski, 1986) and are defined by (see also the next Section)

$$\mathcal{K}_k(\mathbb{L}, Q) := \mathrm{Span}_{\mathbb{R}^1}\left\{Q, \mathbb{L}Q, \ldots, \mathbb{L}^{k-1}Q\right\}, \tag{9}$$

where Q is a fixed observable and $\mathbb{L}$ is a generator of time evolution of the system in question, constitutes a fusion frame in the Hilbert-Schmidt space $\mathcal{B}_*(\mathcal{H})$ if (Jamiołkowski, 2000)

$$\boxplus_{i=1}^{r} \mathcal{K}_\mu(\mathbb{L}, Q_i) = \mathcal{B}_*(\mathcal{H}). \tag{10}$$

In the above equality μ denotes the degree of the minimal polynomial of the superoperator $\mathbb{L}$ and $Q_1, \ldots, Q_r$ represent fixed observables. The symbol $\boxplus$ denotes Minkowski sum of subspaces (10) (see (Hauseholder, 2009; Jamiołkowski, 2010)). We recall that for two subspaces K_1 and K_2 of the vector space $\mathcal{H}$, by $K_1 \boxplus K_2$ one understands the smallest subspace of $\mathcal{H}$ which contains K_1 and K_2.

It is well known that the Krylov subspaces $\mathcal{K}_k(\mathbb{L}, Q)$ for $k = 1, 2, \ldots$ form a nested sequence of subspaces of increasing dimensions that eventually become invariant under $\mathbb{L}$. Hence for a given Q, there exists an index $\mu = \mu(Q)$, often called the *grade of Q with respect to* $\mathbb{L}$ for which

$$\mathcal{K}_1(\mathbb{L}, Q) \subsetneq \cdots \subsetneq \mathcal{K}_\mu(\mathbb{L}, Q) = \mathcal{K}_{\mu+1}(\mathbb{L}, Q) = \mathcal{K}_{\mu+2}(\mathbb{L}, Q) \cdots. \tag{11}$$

It is easy to see, that for a given operator Q, the natural number $\mu(Q)$ is equal to the degree of *the minimal polynomial of* $\mathbb{L}$ *with respect of* Q. Clearly, $\mu(Q) \leq \mu(\mathbb{L})$, where $\mu(\mathbb{L})$ denotes the degree of the minimal polynomial of superoperator $\mathbb{L}$ (cf. e.g. (Jamiołkowski, 2000)). Now, let us observe that even if observables $Q_1, \ldots, Q_r$ are linearly independent, the Krylov subspaces $\mathcal{K}_k(\mathbb{L}, Q_i)$ for $i = 1, \ldots, r$ can have nonempty intersections. At the same time they can constitute a fusion frame for the space of all observables $\mathcal{B}_*(\mathcal{H})$.

In the statistical description of physical systems the main role of observables is to statistically identify states, or some of their properties. A typical goal of an experiment can be to decide among various alternatives or hypothesis about states. As a very good reference on such type of problems we recommend the review book (Paris & Rehacek, 2004). The details of a particular identification problem depend on our prior knowledge and the properties we want to discuss. One can say that owing to both the a priori knowledge about states and the knowledge of our technical possibilities we define the alternatives that we should experimentally verify.

In general, depending whether the set of alternatives is finite or not, one makes a distinction between discrimination and estimation problems. One can introduce three different types of problems:

1. *State estimation problem.* In its most general form, one wants to identify the state of a system assuming that no additional (prior) knowledge is available. In other words, the whole state space of a system constitutes the set of possible hypotheses.
2. *Sufficient statistics for families of states.* In this case we are interested in considering only a subset of the whole set of states. We encode prior knowledge about the preparation of states in a multiparameter family of states and consider them as a possible set of hypotheses. For example, we can assume that one considers states which are pure states or have a particular block-diagonal form.
3. *State discrimination problem.* A particular case of the problem 2). One assumes that we want to identify the state which belongs to a finite set $\{\rho_1, \ldots, \rho_p\}$ and our aim is to distinguish among these p possibilities. It is an obvious observation that in this case the set of observables used for identification can be restricted in an essential way.

All above problems create very interesting particular questions and we will discuss them in separate publications. A general description and some results concerning the problems 2 and 3 based on the idea of fusion frames are discussed in the present paper.

The organization of the paper is as follows: In Section 2, we summarize some concepts and results of the theory of frames; Section 3 presents the main ideas of stroboscopic tomography. We conclude the paper in Section 4 by discussing some applications of the notions of frames and fusion frames to problems of open quantum systems and we discuss some examples of algebraic methods in low-dimensional quantum systems.

2. Frames and fusion frames

Frames were first introduced by Duffin and Schaeffer in 1952 as a natural concept that appeared during their research in nonharmonic Fourier analysis (Duffin & Schaeffer, 1952). After more than three decades Daubechies, Grossman and Meyer (Daubechies et al., 1986) initiated the use of frame theory in the description of signal processing. Today, frame theory plays an important role in dozens of applied areas, cf. e.g. (Christensen, 2008; Heil, 2006; Kovacevic & Chebira, 2008).

Let us consider a Hilbert space $\mathcal{H}$ ($\dim \mathcal{H} = N < \infty$) with scalar product $\langle \cdot | \cdot \rangle$ which is linear in the second argument. A collection of vectors $\mathcal{F} = \{|f_i\rangle : i \in \mathcal{I}\}$, $|f_i\rangle \in \mathcal{H}$, is called *a frame* if there are two positive constants $\alpha, \beta > 0$ such that for every vector $x \in \mathcal{H}$

$$\alpha \parallel x \parallel^2 \leq \sum_{i \in \mathcal{I}} |\langle f_i | x \rangle|^2 \leq \beta \parallel x \parallel^2 . \tag{12}$$

One assumes that the number of vectors $|f_i\rangle$ is greater or equal to N. The frame is *tight* when the constants α and β are equal, $\alpha = \beta$. If $\alpha = \beta = 1$, then $\mathcal{F}$ is called a *Parseval frame.* The numbers $\langle f_i | x \rangle$ are called *frame coefficients.*

For a given frame $\mathcal{F}$ we can introduce the *analysis* Θ and *synthesis* Θ^* operators. They are defined by the equality

$$\Theta(x) = \sum_{i \in \mathcal{I}} \langle f_i | x \rangle |e_i\rangle, \tag{13}$$

where $|e_i\rangle$ stands for the standard basis in $\mathbb{C}^m$ (we will consider only finite dimensional frames, so that $\mathcal{I} = \{1, \ldots, m\}$ and $m \geq N$). Composing Θ with its adjoint operator Θ^*, we obtain the *frame operator*

$$F : \mathcal{H} \to \mathcal{H}, \tag{14}$$

defined by

$$Fx := \Theta^* \Theta x = \sum_{i=1}^{m} \langle f_i | x \rangle |f_i\rangle. \tag{15}$$

It is not difficult to see that any collection of vectors $\{|f_i\rangle\}_{i=1}^m$ constitutes a frame for the vector space $\mathcal{N} := \text{span}\{|f_i\rangle\}_{i=1}^m$, $\mathcal{N} \subseteq \mathcal{H}$. On the other hand a family of elements $\{|f_i\rangle\}_{i=1}^m$ in $\mathcal{H}$ is a frame for $\mathcal{H}$ if and only if $\text{span}\{|f_i\rangle\}_{i=1}^m = \mathcal{H}$. This means that a frame may contain more elements than it is necessary for it to be a basis. In particular, if $\{|f_i\rangle\}_{i=1}^m$ is a frame for $\mathcal{H}$ and $\{|g_i\rangle\}_{i=1}^n$ is an arbitrary finite collection of elements in $\mathcal{H}$, then the set $\{|f_1\rangle, \ldots, |f_m\rangle, |g_1\rangle, \ldots, |g_n\rangle\}$ is also a frame for $\mathcal{H}$.

Generally speaking, frame theory is the study of how $\{|f_i\rangle\}_{i=1}^m$ should be chosen in order to guarantee that the frame operator $\Theta^*\Theta$ is well-conditioned. In particular, $\{|f_i\rangle\}_{i=1}^m$ is a frame for $\mathcal{H}$ if there exist frame bounds α, β such that

$$\alpha\, \mathbb{I} \;\leq\; \Theta^*\Theta \;\leq\; \beta\, \mathbb{I}, \tag{16}$$

and is a tight frame iff $\Theta^*\Theta = \alpha\mathbb{I}$. It is an obvious observation that $F = \Theta^*\Theta$ is a self-adjoint and invertible operator.

Fusion frame theory (theory of frames of subspaces) is an emerging mathematical theory that provides a natural setting for performing distributed data processing in many fields Casazza & Kutyniok (2004); Casazza et al. (2008). In particular, one can apply these ideas in quantum state tomography. The notion of fusion frame was introduced in Casazza & Kutyniok (2004) and further developed by Casazza et al. (2008). A fusion frame in a Hilbert space $\mathcal{H} \cong \mathbb{C}^N$ is a finite collection of subspaces $\{W_i\}_{i=1}^m$ of $\mathcal{H}$, such that there exist constants $0 < \alpha < \beta < \infty$ satisfying, for any $|\varphi\rangle \in \mathcal{H}$, the two inequalities

$$\alpha \parallel |\varphi\rangle \parallel^2 \leq \sum_{i=1}^{m} \parallel \mathbb{P}_i |\varphi\rangle \parallel^2 \leq \beta \parallel |\varphi\rangle \parallel^2, \tag{17}$$

where $\mathbb{P}_i$ denotes the non-orthogonal projection on W_i. In other words, a collection $\{W_i\}_{i=1}^m$ is a fusion frame if and only if

$$\alpha\, \mathbb{I} \;\leq\; \sum_{i=1}^{m} \mathbb{P}_i \;\leq\; \beta\, \mathbb{I}. \tag{18}$$

The constants α and β are called fusion frame bounds. An important class of fusion frames is the class of tight fusion frames, for which $\alpha = \beta$. This equality leads to the operator relation $\sum_{i=1}^m \mathbb{P}_i = \alpha\mathbb{I}$. Let us note that definition given in (Casazza & Kutyniok, 2004; Casazza et al., 2008) for fusion frames applies to weighted subspaces in any Hilbert space as well. However, since the scope of this paper is limited to non-weighted subspaces only, the definition of a fusion frame is presented for this restricted situation. If we compare the definition of a quantum channel and that of a tight fusion frame, it becomes evident that every quantum channel can be considered a special case of a fusion frame (18) with $\alpha = \beta = 1$.

Now, let us recall that for a given operator $M : \mathcal{H} \to \mathcal{H}$ and a given fixed nonzero vector $|x\rangle \in \mathcal{H}$, one introduces the kth-order Krylov subspace of $\mathcal{H}$ by the equality

$$\mathcal{K}_k(M, x) := \text{span}\{|x\rangle, M|x\rangle, \ldots, M^{k-1}|x\rangle\}. \tag{19}$$

The above definition can also be written as

$$\mathcal{K}_k(M, x) := \text{span}\{p(M)|x\rangle;\ \deg(p) \leq k - 1\}, \tag{20}$$

where p denotes an arbitrary polynomial and $\deg(p)$ is its degree. It is an obvious observation that the size of a Krylov subspace depends on both M and $|x\rangle$. Note also that there exists such k that $\mathcal{K}_k(M, x) = \mathcal{K}_{k+1}(M, x)$ and this k is the degree of the minimal polynomial of M with respect to $|x\rangle$. If by $\mu(\lambda, M)$ we denote the minimal polynomial of the operator M, then the minimal polynomial of M with respect to any vector $|x\rangle \in \mathcal{H}$ divides $\mu(\lambda, M)$.

For a given operator $M : \mathcal{H} \to \mathcal{H}$ Krylov subspaces generated by a fixed set of vectors $|x_1\rangle, \ldots, |x_r\rangle$ constitute a fusion frame in $\mathcal{H}$ if and only if the following equality is satisfied

$$\boxplus_{i=1}^{r} \mathcal{K}_\mu(M, x_i) = \mathcal{H}. \tag{21}$$

3. Stroboscopic tomography of open quantum systems

Quantum theory — as a description of properties of microsystems — was born more then a hundred years ago. But for a long time it was merely a theory of isolated systems. Only around fifty years ago the theory of quantum systems was generalized. The so-called *theory of open quantum systems* (systems interacting with their environments) was established, and the main sources of inspiration for it were quantum optics and the theory of lasers. This led to the generalization of states (now density operators are considered to be a natural representation of quantum states), and to generalized description of their time evolution. At that time the concept of so-called *quantum master equations* — which preserve positive semi-definiteness of density operators — and the idea of a *quantum communication channel* were born, cf. e.g. (Gorini et al., 1976; Kossakowski, 1972; Kraus, 1971; Lindblad, 1976). On the mathematical level, this approach initiated the study of semigroups of completely positive maps and their generators. Now, for the convenience of the readers, we summarize the main ideas and methods of description of open quantum systems and the so-called stroboscopic tomography.

The time evolution of a quantum system of finitely many degrees of freedom (a qudit) coupled with an infinite quantum system, usually called a reservoir, can be described, under certain limiting conditions, by a one-parameter semigroup of maps (cf. e.g. (Gorini et al., 1976; Jamiołkowski, 1974; Kossakowski, 1972)). Let $\mathcal{H}$ be the Hilbert space of the first system ($\dim \mathcal{H} = N$) and let

$$\Phi(t) : \mathcal{B}_\star(\mathcal{H}) \to \mathcal{B}_\star(\mathcal{H}), \quad t \in \mathbb{R}^1_+, \tag{22}$$

be a dynamical semigroup, where $\mathcal{B}_\star(\mathcal{H})$ denotes the real vector space of all self-adjoint operators on $\mathcal{H}$. If one introduces the scalar product of operators A, B by the formula $\langle A, B\rangle = \text{Tr}(A^* B)$, then $\mathcal{B}_\star(\mathcal{H})$ can be considered as yet another inner product space, namely the so-called Hilbert-Schmidt space with the norm defined by $\| \rho \|^2 = \text{Tr}(\rho^* \rho)$. States of the system are described by density operators $\rho \in S(\mathcal{H})$, where

$$S(\mathcal{H}) := \{\rho \in \mathcal{B}_\star(\mathcal{H});\ \rho \geq 0, \mathrm{Tr}\,\rho = 1\}\,. \tag{23}$$

Usually one assumes that the family of linear superoperators $\Phi(t)$ satisfies

1. $\Phi(t)$ is trace preserving, $t \in \mathbb{R}^1_+$,
2. $\| \Phi(t)\rho \| \leq \| \rho \|$ for all $\rho \in \mathcal{B}_\star(\mathcal{H})$,
3. $\Phi(t_1) \circ \Phi(t_2) = \Phi(t_1 + t_2)$,

for all t_1, t_2 in $\mathbb{R}^1_+$, and if $t \to 0$, then $\lim \Phi(t) = \mathbb{I}$. Since such defined $\Phi(t)$ is a contraction, it follows from the Hille-Yosida theorem that there exists a linear superoperator $\mathbb{K} : \mathcal{B}_\star(\mathcal{H}) \to \mathcal{B}_\star(\mathcal{H})$ such that $\Phi(t) = \exp(t\mathbb{K})$ for all $t \geq 0$ and

$$\frac{\mathrm{d}\rho(t)}{\mathrm{d}t} = \mathbb{K}\rho(t), \tag{24}$$

where $\rho(t) = \Phi(t)\rho(0)$. One should stress that the above conditions for semigroup $\Phi(t)$ imply preservation of positivity of density operators, $\rho(0) \geq 0 \Rightarrow \rho(t) = \Phi(t)\rho(0) \geq 0$ for all $t \in \mathbb{R}^1_+$. Now, the above equation (usually called the *master equation*) defines an assignment (the trajectory of $\rho(0)$)

$$\mathbb{R}^1_+ \ni t \mapsto \rho(t) \in S(\mathcal{H}), \tag{25}$$

provided that we know the initial state of the system $\rho(0) \in S(\mathcal{H})$. The fundamental question of the stroboscopic tomography reads: What can we say about the trajectories (initial state $\rho(0)$) if the only information about the system in question is given by the mean values

$$E_i(t_j) = \mathrm{Tr}\,(Q_i\rho(t_j)), \tag{26}$$

of, say, r linearly independent self-adjoint operators $Q_1, \ldots, Q_r$ at some instants $t_1, \ldots, t_p$, where $r < N^2 - 1$ and $t_j \in [0, T]$ for $j = 1, \ldots, p$, $T > 0$. In other words, the problem of the stroboscopic tomography consists in the reconstruction of the initial state $\rho(0)$, or a current state $\rho(t)$ for any $t \in \mathbb{R}^1_+$, from known expectation values (26). To be more precise we introduce the following description. Suppose that we can prepare a quantum system repeatedly in the same initial state and we make a series of experiments such that we know the expectation values $E_Q(t_j) = \mathrm{Tr}\,(Q\rho(t_j))$ for a fixed set of observables $Q_1, \ldots, Q_r$ at different time instants $t_1 < t_2 < \cdots < t_p$. The basic question is: can we find the expectation value of any other operator $Q \in \mathcal{B}_\star(\mathcal{H})$, that is any other observable from $\mathcal{B}_\star(\mathcal{H})$, knowing the set of measured outcomes of a given set $Q_1, \ldots, Q_r$ at $t_1, \ldots, t_p$, i.e. knowing $E_j(t_k)$ for $j = 1, \ldots, r$ and where $0 \leq t_1 < t_2 < \cdots < t_p \leq T$, for an interval $[0, T]$?

If the problem under consideration is static, then the state of a N-level open quantum system (a qudit) can be uniquely determined only if $r = N^2 - 1$ expectation values of linearly independent observables are at our disposal. However, if we assume that we know the dynamics of our system i.e. we know the generator $\mathbb{K}$ or $\mathbb{L} := (\mathbb{K})^*$ (in the Heisenberg picture) of the time evolution, then we can use the stroboscopic approach based on a discrete set of times $t_1, ..., t_p$. In general, we use the term "state-tomography" to denote any kind of state-reconstruction method.

With reference to the terminology used in system theory, we introduce the following definition: An N-level open quantum system $\mathbb{S}$ is said to be $(Q_1, \ldots, Q_r)$-reconstructible on the interval $[0, T]$, if for every two trajectories defined by the equation (24) there exists at least one instant $\hat{t} \in [0, T]$ and at least one operator $Q_k \in \{Q_1, \ldots, Q_r\}$ such that

$$\mathrm{Tr}\,(Q_k \rho_1(\hat{t})) \neq \mathrm{Tr}\,(Q_k \rho_2(\hat{t})). \tag{27}$$

The above definition is equivalent to the following statement. An N-level open quantum system $\mathbb{S}$ is $(Q_1, \ldots, Q_r)$-reconstructible on the interval $[0, T]$ iff there exists at least one set of time instants $0 < t_1 < \cdots < t_p \leq T$ such that the state trajectory can be uniquely determined by the correspondence

$$[0, T] \ni t_j \longmapsto E_i(t_j) = \mathrm{Tr}\,(Q_i \rho(t_j)), \tag{28}$$

for $i = 1, \ldots, r$ and $j = 1, \ldots, p$.

Let us observe that in the above definition of reconstructibility we discuss the problem of verifying whether the accessible information about the system is sufficient to determine the state uniquely and we do not insist on determining it explicitly.

The positive dynamical semigroup $\{\Phi(t),\ t \in \mathbb{R}^1_+\}$ is determined by the generator $\mathbb{K} : \mathcal{B}_\star(\mathcal{H}) \to \mathcal{B}_\star(\mathcal{H})$ (the Schrödinger picture) and it is related to the generator $\mathbb{L}$ of the semigroup in the Heisenberg picture by the duality relation

$$\mathrm{Tr}[Q(\mathbb{K}\rho)] = \mathrm{Tr}[(\mathbb{L}Q)\rho]. \tag{29}$$

For a given set of observables $Q_1, \ldots, Q_r$, the subspace spanned on the operators

$$Q_i, \mathbb{L}Q_i, \ldots, (\mathbb{L})^{k-1}Q_i,$$

will be denoted by

$$\mathcal{K}_k(\mathbb{L}, Q_i) := \mathrm{Span}_{\mathbb{R}^1}\left\{Q_i, \mathbb{L}Q_i, \ldots, \mathbb{L}^{k-1}Q_i\right\}, \tag{30}$$

as the Krylov subspace in the Hilbert-Schmidt space $\mathcal{B}_\star(\mathcal{H})$. If $k = \mu$, where μ is the degree of the minimal polynomial of the generator $\mathbb{L}$, then the subspace $\mathcal{K}_\mu(\mathbb{L}, Q_i)$ is an invariant subspace of the superoperator $\mathbb{L}$ with respect to Q_i. It can be easily seen that the subspace $\mathcal{K}_\mu(\mathbb{L}, Q_i)$ is essentially spanned on all operators of the form $(\mathbb{L})^k Q_i$, where $k = 0, 1, \ldots$. Furthermore, it is the smallest invariant subspace of the superoperator $\mathbb{L}$ containing Q_i (i.e. the common part of all invariant subspaces of the operator $\mathbb{L}$ containing Q_i).

One can now formulate the sufficient conditions for the reconstructibility of an N-level open quantum system (c.f. Jamiołkowski (1983; 2000)).

Let $\mathbb{S}$ be an N-level open quantum system with the evolution governed by an equation of the form $\dot{Q}(t) = \mathbb{L}Q(t)$ (the Heisenberg picture), where $\mathbb{L}$ is the generator of the dynamical semigroup $\Psi(t) = \exp(t\mathbb{L})$. Suppose that, by performing measurements, the correspondence

$$[0, T] \ni t_j \longmapsto E_i(t_j) = \mathrm{Tr}\,(\rho(0) Q_i(t_j)) \tag{31}$$

can be established for fixed observables $Q_1, \dots, Q_r$ at selected time instants $t_1, \dots, t_p$. The system $\mathbb{S}$ is $(Q_1, \dots, Q_r)$-reconstructible if

$$\boxplus_{i=1}^{r} \mathcal{K}_\mu(\mathbb{L}, Q_i) = \mathcal{B}_\star(\mathcal{H}). \tag{32}$$

The above condition has been obtained by using the polynomial representation of the semigroup $\Psi(t)$. Indeed, if $\mu(\lambda, \mathbb{L})$ denotes the minimal polynomial of the generator $\mathbb{L}$ and $\mu = \deg \mu(\lambda, \mathbb{L})$, then $\Psi(t) = \exp(t\mathbb{L})$ can be represented in the form

$$\Psi(t) = \sum_{k=0}^{\mu-1} \alpha_k(t) \mathbb{L}^k, \tag{33}$$

where the functions $\alpha_k(t)$ for $k = 0, \dots, \mu - 1$ are particular solutions of the scalar linear differential equation with *characteristic polynomial* $\mu(\lambda, \mathbb{L})$. Since the functions $\alpha_k(t)$ are mutually independent, therefore for arbitrary $T > 0$ there exists at least one set of moments $t_1, \dots, t_\mu$ ($\mu = \deg \mu(\lambda, \mathbb{L})$) such that

$$0 \leq t_1 < t_2 < \cdots < t_\mu \leq T, \tag{34}$$

and $\det[\alpha_k(t_j)] \neq 0$. Taking into account these conditions one finds that the state $\rho(0)$ can be determined uniquely if operators of the form

$$f_{kl} := (\mathbb{L})^k Q_l \tag{35}$$

for $l = 1, \dots, r$ and $k = 0, 1, \dots$ span the space $\mathcal{B}_\star(\mathcal{H})$. In other words, we can say that $\rho(0)$ can be determined if vectors (35) constitute a frame in Hilbert-Schmidt space $\mathcal{B}_\star(\mathcal{H})$ or, equivalently, if Krylov subspaces $\mathcal{K}_\mu(\mathbb{L}, Q_l)$ for $l = 1, \dots, r$ constitute a fusion frame in $\mathcal{B}_\star(\mathcal{H})$.

It should be noted that almost all the above considerations can be generalized to infinite dimensional Hilbert spaces (Lindblad, 1976, Jamiołkowski, 1982). Such approach is also discussed in a recent literature on infinite dimensional Kraus operators describing amplitude-damping channels and laser processes. For instance, the above techniques are used in the description of such situations in which beamsplitters allow photons to be coupled to another optical modes representing the environment (cf. e.g. Fan & Hu).

3.1 Minimal number of observables

The question of an obvious physical interest is to find the minimal number of observables $Q_1, \dots, Q_\eta$ for which an N-level quantum system $\mathbb{S}$ with a fixed generator $\mathbb{L}$ can be $(Q_1, \dots, Q_\eta)$-reconstructible. It can be shown that for an N-level generator there always exists a set of observables $Q_1, \dots, Q_\eta$, where

$$\eta := \max_{\lambda \in \sigma(\mathbb{L})} \{\dim \operatorname{Ker}(\lambda \mathbb{I} - \mathbb{L})\}, \tag{36}$$

such that the system is $(Q_1, \dots, Q_\eta)$-reconstructible (Jamiołkowski, 2000). Moreover, if we have another set of observables $\widetilde{Q}_1, \dots, \widetilde{Q}_{\widetilde{\eta}}$ such that the system is $(\widetilde{Q}_1, \dots, \widetilde{Q}_{\widetilde{\eta}})$-reconstructible, then $\widetilde{\eta} \geq \eta$. The number η defined by (36) is called the *index of cyclicity* of the quantum open system $\mathbb{S}$ (Jamiołkowski, 2000). The symbol $\sigma(\mathbb{L})$ in (36) denotes the spectrum of the superoperator $\mathbb{L}$.

In particular, if we consider an isolated quantum system characterized by Hamiltonian H_0, then the minimal number of observables $Q_1, \ldots, Q_\eta$ for which the system is $(Q_1, \ldots, Q_\eta)$-reconstructible is given by

$$\eta = n_1^2 + n_2^2 + \cdots + n_m^2, \tag{37}$$

where $n_i = \dim \operatorname{Ker}(\lambda_i I - H_0)$ for all $\lambda_i \in \sigma(H_0)$, $i = 1, \ldots, m$ (for details cf. Jamiołkowski (1982; 2000)).

Now let us assume that the time evolution of an N-level quantum system $\mathcal{S}$ is described by the generator $\mathbb{L}$ given by

$$\mathbb{L}\rho = \frac{1}{2}\{[R\rho, R] + [R, \rho R]\} = -\frac{1}{2}[R, [R, \rho]], \tag{38}$$

that is, we consider the so-called Gaussian semigroup. The symbol R in (38) denotes a self-adjoint operator with the spectrum

$$\sigma(R) = \{\lambda_1, \ldots, \lambda_m\}. \tag{39}$$

In the sequel n_i stands for the multiplicity of the eigenvalue λ_i for $i = 1, \ldots, m$. One can assume that the elements of the spectrum of R are numbered in such a way that the inequalities $\lambda_1 < \lambda_2 < \ldots < \lambda_m$ are fulfilled. The following theorem holds:

The index of cyclicity of the Gaussian semigroup with a generator $\mathbb{L}$ given by (38) is expressed by the formula

$$\eta = \max\{\kappa, \gamma_1, \ldots, \gamma_r\}, \tag{40}$$

where $r = (m-1)/2$ if m is odd or $r = (m-2)/2$ if m is even, and

$$\kappa := n_1^2 + n_2^2 + \ldots + n_m^2, \tag{41}$$

$$\gamma_k := 2\sum_{i=1}^{m-k} n_i\, n_{i+k}. \tag{42}$$

In order to prove the above theorem and to determine the value of η for the generator $\mathbb{L}$ defined by (38) we must find the number of nontrivial invariant factors of the operator $\mathbb{L}$. Let us observe that if $\sigma(N) = \{\lambda_1, \ldots, \lambda_m\}$, then the spectrum of the operator $\mathbb{L}$ is given by

$$\sigma(\mathbb{L}) = \Big\{\nu_{ij} \in \mathbb{R};\ \nu_{ij} = (\lambda_i - \lambda_j)^2,\ i, j = 1, \ldots, m\Big\}. \tag{43}$$

The above statement follows from the fact that the operator $\mathbb{L}$ can also be represented as

$$\mathbb{L} = R^2 \otimes I + I \otimes R^2 - 2R \otimes R, \tag{44}$$

where I denotes the identity in the space $\mathcal{H}$. Since R is self-adjoint therefore the algebraic multiplicity of λ_i, i.e. the multiplicity of λ_i as the root of the characteristic polynomial of R, is equal to the geometric multiplicity of λ_i, $n_i = \dim \text{Ker}\,(\lambda_i I - R)$. Of course, we have $n_1 + \ldots + n_m = \dim \mathcal{H}$.

From (44) we can see that the multiplicities of the eigenvalues of the operator $\mathbb{L}$ are not determined uniquely by the multiplicities of $\lambda_i \in \sigma(R)$. But if we assume that $\lambda_1 < \ldots < \lambda_m$ and $\lambda_k = (k-1)c + \lambda_1$, where $k = 1, \ldots, m$, and $c = \text{const} > 0$, then the multiplicities of all eigenvalues of $\mathbb{L}$ are given by

$$\gamma_{|i-j|} = \dim \text{Ker}\,[(\lambda_i - \lambda_j)^2 \mathbb{I} - \mathbb{L}] \tag{45}$$

for $i \neq j$ and

$$\dim \text{Ker}\,(\mathbb{L}) = n_1^2 + \ldots + n_m^2 = \kappa \tag{46}$$

when $i = j$. Now, as we know, the minimal number of observables $Q_1, \ldots, Q_\eta$ for which the qudit $\mathcal{S}$ can be $(Q_1, \ldots, Q_\eta)$-reconstructible is given by (36), so in our case

$$\eta = \max_{i,j=1,\ldots,m} \left\{ \dim \text{Ker}\,[(\lambda_i - \lambda_j)^2 \mathbb{I} - \mathbb{L}] \right\}, \tag{47}$$

where $\lambda_i \in \sigma(R)$. Using the above formulae and the inequality $\gamma_k < \kappa$ for $k > r$, where r is given by $(m-1)/2$ if m is odd and $(m-2)/2$ if m is even, we can observe that also without the assumption $\lambda_k = (k-1)c + \lambda_1$ one obtains

$$\eta = \max\{\kappa, \gamma_1, \ldots, \gamma_r\}. \tag{48}$$

This completes the proof.

3.2 The choice of moments of observations

Another natural question arises: what are the criteria governing the choice of time instants $t_1, \ldots, t_\mu$? The following theorem holds:

Let us assume that $0 \leq t_1 < t_2 < \ldots < t_\mu \leq T$. Suppose that the mutual distribution of time instants $t_1, \ldots, t_\mu$ is fixed, i.e. a set of nonnegative numbers $c_1 < \ldots < c_\mu$ is given and $t_j := c_j t$ for $j = 1, \ldots, \mu$, and $t \in \mathbb{R}_+$. Then for $T > 0$ the set

$$\tau(T) := \left\{ (t_1, \ldots, t_\mu) : \; t_j = c_j t, \; 0 \leq t \leq \frac{T}{c_\mu} \right\}$$

contains almost all sequences of time instants $t_1, \ldots, t_\mu$, i.e. all of them except a finite number.

As one can check, the expectation values $E_i(t_j)$ and the operators $(\mathbb{L})^k Q_i$ are related by the equality

$$E_i(t_j) = \sum_{k=0}^{\mu-1} \alpha_k(c_j t) \Big((\mathbb{L})^k Q_i, \rho_0 \Big), \tag{49}$$

where we assume that $t_j = c_j t$ and the bracket $(\cdot, \cdot)$ denotes the Hilbert-Schmidt product in $\mathcal{B}_*(\mathcal{H})$. One can determine ρ_0 from (49) for all those values $t \in \mathbb{R}_+$ for which the determinant $\Omega(t)$ is different from zero, i.e.

$$\Omega(t) := \det\,[\alpha_k(c_j t)] \neq 0. \tag{50}$$

One can prove that the range of the parameter $t \in \mathbb{R}_+$ for which $\Omega(t) = 0$ consists only of isolated points on the semiaxis $\mathbb{R}_+$, i.e. does not possess any accumulation points on $\mathbb{R}_+$. To this end let us note that since the functions $t \to \alpha_k(t)$ for $k = 0, 1, \ldots, \mu - 1$, are analytic on $\mathbb{R}$, the determinant $\Omega(t)$ defined by (50) is also an analytic function of $t \in \mathbb{R}$. If $\Omega(t)$ can be proved to be nonvanishing identically on $\mathbb{R}$, then, making use of its analyticity, we shall be in position to conclude that the values of t, for which $\Omega(t) = 0$, are isolated points on the axis $\mathbb{R}$.

It is easy to check that for $k = \mu(\mu - 1)/2$

$$\left.\frac{d^k\Omega(t)}{dt^k}\right|_{t=0} = \prod_{1 \leq j < i \leq \mu} (c_i - c_j) . \tag{51}$$

According to the assumption $c_1 < c_2 < \ldots < c_\mu$, we have $\Omega^{(k)}(0) \neq 0$ if $k = \mu(\mu - 1)/2$. This means that the analytic function $t \to \Omega(t)$ does not vanish identically on $\mathbb{R}$ and the set of values of t for which $\Omega(t) = 0$ cannot contain accumulation points. In other words, if we limit ourselves to an arbitrary finite interval $[0, T]$, then $\Omega(t)$ can vanish only on a finite number of points belonging to $[0, T]$. This completes the proof.

4. Frames and fusion frames in stroboscopic tomography. Generalizations to subalgebras

As we have seen the concepts of frames and fusion frames appear in stroboscopic tomography in natural way. The conclusion is based on the discussed above polynomial representations of semigroups which describe evolutions of open systems. The possibility to represent the semigroup $\Phi(t) = \exp(t\,\mathbb{L})$ in the form

$$\Phi(t) = \sum_{k=0}^{\mu-1} \alpha_k(t)\mathbb{L}^k, \tag{52}$$

where μ stands for the degree of the minimal polynomial of the superoperator $\mathbb{L}$ and $\alpha_k(t)$, $k = 0 \ldots, \mu - 1$, denote some functions of the eigenvalues of $\mathbb{L}$ gives the equality (32) as a sufficient condition for stroboscopic tomography. On the other hand, this equality means that the Krylov subspaces $\mathcal{K}_\mu(\mathbb{L}, Q_i)$, $i = 1, \ldots, r$, constitute a fusion frame in the Hilbert-Schmidt space $\mathcal{B}_*(\mathcal{H})$ of all observables. Moreover, this also means that the collection of vectors

$$f_{jk} := \mathbb{L}^k Q_j, \tag{53}$$

for $j = 1, \ldots, r$ and $k = 0, 1, \ldots, \mu - 1$, constitute a frame in $\mathcal{B}_*(\mathcal{H})$ and the system in question is $(Q_1, \ldots, Q_r)$-reconstructible. In this case every element Q of the space $\mathcal{B}_*(\mathcal{H})$ can be represented as

$$Q = \sum_{j,k} \langle F^{-1} f_{jk} | Q \rangle f_{jk} = \sum_{j,k} \langle f_{jk} | Q \rangle F^{-1} f_{jk}, \tag{54}$$

where F denotes the frame operator of the collection of vectors (53). One can say even more. If $Q \in \mathcal{B}_*(\mathcal{H})$ also has another representation $Q = \sum_{j,k} c_{jk} f_{jk}$ for some scalar coefficients c_{jk},

$j = 1, \ldots, r$ and $k = 0, 1, \ldots, \mu - 1$, then

$$\sum_{jk} |c_{jk}|^2 = \sum_{jk} |\langle F^{-1} f_{jk} | Q \rangle|^2 + \sum_{j,k} |c_{jk} - \langle F^{-1} f_{jk} | Q \rangle|^2. \tag{55}$$

It is obvious that every frame in finite-dimensional space contains a subset that is a basis. As a conclusion we can say that if $\{f_{jk}\}$ is a frame but not a basis, then there exists a set of scalars $\{d_{jk}\}$ such that $\sum_{j,k} d_{jk} f_{jk} = 0$. Therefore, any fixed element Q of $\mathcal{B}_*(\mathcal{H})$ can also be represented as

$$Q = \sum_{j,k} \left(\langle F^{-1} f_{jk} | Q \rangle + d_{jk} \right) f_{jk}. \tag{56}$$

The above equality means that every $Q \in \mathcal{B}_*(\mathcal{H})$ has many representations as superpositions of elements from the set (53). But according to equality (55) among all scalar coefficients $\{c_{jk}\}$ for which

$$Q = \sum_{j,k} c_{jk} f_{jk}, \tag{57}$$

the sequence $\{\langle F^{-1} f_{jk} | Q \rangle\}$ has minimal norm. This is a general method in frame theory (Christensen, 2008) and at the same time the main observation connected with the idea of stroboscopic tomography.

In conclusion, one can say that the Krylov subspaces $\mathcal{K}_\mu(\mathbb{L}, Q_i)$ in the space $\mathcal{B}_*(\mathcal{H})$ generated by the superoperator $\mathbb{L}$ can be used in an effective way for procedures of stroboscopic tomography if they constitute appropriate fusion frames in this space.

4.1 Generalizations to subalgebras

Now, we will discuss some problems of reconstruction of quantum states when the Krylov subspaces playing such important role in the stroboscopic tomography are replaced by some subalgebras of the Hilbert-Schmidt space $\mathcal{B}_\star(\mathcal{H})$. Just as the fundamental theorem of algebra ensures that every linear operator acting on a finite dimensional complex Hilbert space has a nontrivial invariant subspace, the *fundamental theorem of noncommutative algebra* asserts the existence of invariant subspaces of $\mathcal{H}$ for some families of operators from $\mathcal{B}(\mathcal{H})$. It is an obvious observation that an algebra generated by any fixed operator Q and the identity on $\mathcal{H}$ can not be equal to $\mathcal{B}_\star(\mathcal{H})$. This statement is based on the Hamilton-Cayley theorem. However, already for two operators Q_1, Q_2 and the identity we can have $\mathrm{Alg}(I, Q_1, Q_2)$ $=\mathcal{B}(\mathcal{H})$ (for details cf. below).

In general, the famous Burnside's theorem states (cf. e.g. (Farenick, 2001)) that an operator algebra on a finite-dimensional vector space with no nontrivial subspaces must be the algebra of *all* linear operators. In the sequel we will use the following version of this theorem:

Fundamental theorem of noncommutative algebras. If $\mathcal{A}$ is a proper subalgebra of $\mathcal{B}(\mathcal{H})$ containing identity, and the dimension of the Hilbert space $\mathcal{H}$ is greater or equal to 2, then $\mathcal{A}$ has a proper nonzero invariant subspace in $\mathcal{H}$ (i.e., the subspace is invariant for all members Q of the algebra $\mathcal{A}$).

We will apply the above theorem for the following problem. Given a set $\mathcal{F} = \{Q_1, \ldots, Q_r\}$ of observables, we would like to establish conditions, when the operators $Q_1, \ldots, Q_r$ generate the whole algebra $\mathcal{B}(\mathcal{H})$. In other words, we want to determine whether every element in $\mathcal{B}(\mathcal{H})$ can be represented in the form $\pi(Q_1, \ldots, Q_r)$, where π is a polynomial in noncommutative variables.

Let us observe that according to the *fundamental theorem* if $\mathcal{A}$ is a subalgebra of the full complex algebra $\mathcal{B}(\mathcal{H})$, then a nontrivial invariant subspace in $\mathcal{H}$ exists if and only if

$$\dim \mathcal{A} < \dim \mathcal{B}(\mathcal{H}). \tag{58}$$

If a set of generators of $\mathcal{A}$ is known, then the above inequality can be verified by a finite number of arithmetic operations. The procedures possessing such property are called *effective*. A very important example of an effective procedure can be formulated when we discuss the problem of the existence a common one-dimensional invariant subspace for a pair of operators Q_1, Q_2. In other words, we ask about a common eigenvector for two operators Q_1, Q_2. An answer to this question is given by the following procedure. Let the symbol $[Q_1, Q_2]$ denote, as usual, the commutator of the operators Q_1, Q_2. Then a common eigenvector for Q_1 and Q_2 exists if and only if the subspace $\mathcal{K}$ of $\mathcal{H}$ defined by

$$\mathcal{K} := \bigcap_{\substack{j=1\\k=1}}^{N-1} \operatorname{Ker}[Q_1^j, Q_2^k], \tag{59}$$

where $N = \dim \mathcal{H}$, satisfies the condition $\dim \mathcal{K} > 0$ (this is the so-called Shemesh criterion (Shemesh, 1984)). A short proof of this condition is possible.

First of all, let us observe that if $|\psi\rangle$ is a common eigenvector of the operators Q_1 and Q_2, i.e.,

$$Q_1|\psi\rangle = \alpha|\psi\rangle \quad \text{and} \quad Q_2|\psi\rangle = \beta|\psi\rangle, \tag{60}$$

then $|\psi\rangle$ belongs to $\operatorname{Ker}[Q_1^j, Q_2^k]$ for all j, k greater then 1. This fact and the inequality $\dim \mathcal{K} > 0$ means that the gist of the Shemesh condition is in observation that the subspace $\mathcal{K}$ is invariant under Q_1 and Q_2. Indeed, if $|\psi\rangle$ belongs to $\mathcal{K}$, then by the definition of subspaces $\operatorname{Ker}[Q_1^j, Q_2^k]$ one can check that $Q_1|\psi\rangle \in \mathcal{K}$ and $Q_2|\psi\rangle \in \mathcal{K}$. Now, let us choose a basis for $\mathcal{K}$ and extend it to a basis in $\mathcal{H}$. We then observe that there exists a nonsingular matrix S such that matrices SQ_1S^{-1} and SQ_2S^{-1} have block-triangular forms and the submatrices which correspond to subspace $\mathcal{K}$ commute. This means that these submatrices have a common eigenvector and therefore the same is true for Q_1 and Q_2. D. Shemesh observed that the condition $\dim \mathcal{K} > 0$ is equivalent to the singularity of the matrix

$$\mathbb{M} := \sum_{\substack{j=1\\k=1}}^{N-1} [Q_1^j, Q_2^k]^* [Q_1^j, Q_2^k], \tag{61}$$

where * denotes complex conjugate transpose. For our purposes, on the basis of Burnside's theorem, more interesting is the case when matrices Q_1, Q_2 do not have common eigenvectors and the algebra $A(Q_1, Q_2)$ generated by them coincides with $\mathcal{B}(\mathcal{H})$. This situation may be expressed by the following inequality

$$\det \mathbb{M} > 0, \tag{62}$$

which can be checked by an effective procedure, that is, by a finite number of arithmetic operations. It is obvious, that the matrix $\mathbb{M}$ is in general semipositive definite, and the above condition means the strict positivity of $\mathbb{M}$.

4.2 Examples

In order to illustrate algebraic methods in reconstruction problems, we will discuss some algebraic procedures in low dimensional cases. For quantum systems of qubits and qutrits one can formulate an explicit form of some conditions in a matrix form which is sometimes more transparent then the general operator form. We will use the so-called *vec operator* procedure which transforms a matrix into a vector by stacking its columns one underneath the other. It is well known, that the tensor product of matrices and the vec operator are intimately connected. If A denotes a $N \times N$ matrix and a_j its j-th column, then vec A is the N^2-dimensional vector constructed from $a_1, \ldots, a_N$. Moreover if A, B, C are three matrices such that the matrix product ABC is well defined, then

$$\mathrm{vec}(ABC) = (C^T \otimes A)\,\mathrm{vec}\,B. \tag{63}$$

In the above formula C^T denotes the transposition of the matrix C. In particular we have

$$\mathrm{vec}\,A = (\mathbb{I} \otimes A)\,\mathrm{vec}\,\mathbb{I} = (A^T \otimes \mathbb{I})\,\mathrm{vec}\,\mathbb{I}. \tag{64}$$

Let us agree that when we say that a set of matrices generates the set $\mathcal{B}(\mathcal{H})$, we are thinking about $\mathcal{B}(\mathcal{H})$ as an algebra, while when we say that a set of matrices forms a basis for $\mathcal{B}(\mathcal{H})$, we are talking about $\mathcal{B}(\mathcal{H})$ as a vector space (here we identify $\mathcal{B}(\mathcal{H})$ with the set of all matrices on $\mathcal{H} = \mathbb{C}^N$).

For qubits, that is for two-dimensional Hilbert space, one can show by a direct computation that

$$\det(\mathrm{vec}\,\mathbb{I}, \mathrm{vec}\,Q_1, \mathrm{vec}\,Q_2, \mathrm{vec}(Q_1Q_2)) = \det([Q_1, Q_2]) \tag{65}$$

and

$$\det(\mathrm{vec}\,\mathbb{I}, \mathrm{vec}\,Q_1, \mathrm{vec}\,Q_2, \mathrm{vec}[Q_1, Q_2]) = 2\det([Q_1, Q_2]), \tag{66}$$

where on the left hand side we have the determinants of the 4×4 matrices and on the right hand sides $[Q_1, Q_2]$ denotes the commutator of the two 2×2 matrices.

From the last equality it follows, that if matrices $\mathbb{I}, Q_1, Q_2$ and $[Q_1, Q_2]$ are linearly independent, then the algebra which is spanned by them has the dimension 4, so Q_1, Q_2 and $\mathbb{I}$ generate $\mathcal{B}(\mathcal{H})$. In other words, two operators Q_1, Q_2 and the identity generate $\mathcal{B}(\mathcal{H})$ if and only if the matrix $[Q_1, Q_2]$ has the determinant different from zero. In a similar way one can show that the matrices Q_1, Q_2, Q_3, such that *no* two of them generate $\mathcal{B}(\mathcal{H})$, can generate $\mathcal{B}(\mathcal{H})$ if and only if the double commutator $[Q_1, [Q_2, Q_3]]$ is invertible. In general, the matrices $Q_1, \ldots, Q_r$ generate $\mathcal{B}(\mathcal{H})$ iff at least one of the commutators $[Q_i, Q_j]$ or double commutators $[Q_i, [Q_j, Q_k]]$ is invertible (Aslaksen & Sletsjøe, 2009).

In the case of qutrits, that is for a three-dimensional Hilbert space, one can show by direct calculation that if $[Q_1, Q_2]$ is invertible and $\omega([Q_1, Q_2]) \neq 0$, where for $Q \in \mathcal{B}(\mathcal{H})$ the symbol $\omega(Q)$ denotes the linear term in the characteristic polynomial of Q, then one can construct an explicit basis for $\mathcal{B}(\mathcal{H})$. Indeed, if Q_1, Q_2 belong to $\mathcal{B}(\mathcal{H})$, and $(\dim \mathcal{H}) = 3$, then the determinant of the 9-dimensional matrix Ω build from vec transformations of $\mathbb{I}, Q_1, Q_2, Q_1^2, Q_2^2, Q_1Q_2, Q_2Q_1, [Q_1, [Q_1, Q_2]], [Q_2, [Q_2, Q_1]]$ satisfies the equality

$$\det \Omega = 9 \det([Q_1, Q_2]) \omega([Q_1, Q_2]). \tag{67}$$

That is, if $\det([Q_1, Q_2]) \neq 0$ and $\omega(Q) \neq 0$, then the columns of the matrix Ω correspond to a basis for $\mathcal{B}(\mathcal{H})$.

Of course, one can also use the Shemesh criterion to characterize pairs of generators for $\mathcal{B}(\mathcal{H})$, where $\dim \mathcal{H} = 3$.

5. Conclusions

Papers written by mathematicians are usually focused on characterization of various properties of discussed objects and search for necessary and sufficient conditions for desired conclusion to hold. Concrete constructions offen play a minor role. The problems of frames and fusion frames are no exceptions. The main purpose of this paper was to discuss properties of some Krylov subspaces in a given Hilbert space as a natural examples of fusion frames and their applications in reconstruction of trajectories of open quantum systems.

6. References

Aslaksen, H. & Sletsjøe, A.B. (2009). *Lin. Algebra Appl.* 430, 1.

Casazza, P. G. & Kutyniok, G. (2004). Frames of subspaces, *Contemp. Math.* 345, 87.

Casazza, P. G. et al. (2008). *Appl. Comput. Harmon. Anal.* 25, 114.

Christensen, O. (2008). *Frames and Bases*, Boston, Birkhäuser.

Daubechies, I., Grossman. A., and Meyer, Y. (1986). *J. Math. Phys.* 27, 1271.

Duffin, R. J. & Schaeffer, A. C. (1952). *Transactions AMS* 72, 341.

Fan, H., Hu, L. (2009) *Optics Comm.* 282, 932.

Farenick, D.R. (2001), *Algebras of Linear Transformations*, New York, Springer.

Gorini, V et al. (1976) *J. Math. Phys.* 17, 149.

Hauseholder, A.S. (2009). *The Theory of Matrices in Numerical Analysis*, Dover Books.

Heil, Ch. (Ed.) (2006). *Harmonic Analysis and Applications*, Boston, Birkhäuser.

Jamiołkowski, A. (1974). *Rep. Math. Phys.* 5, 415.

Jamiołkowski, A. (1982). *On Observability of Classical and Quantum Stochastic Systems*, Toruń, N. Copernicus Univ. Press, (in Polish).

Jamiołkowski, A. (1983). *Internat. J. Theoret. Phys.* 22, 369.

Jamiołkowski, A. (1986). *Frequency Domain and State Space Methods for Linear Systems*, Eds. C. I. Byrnes and A. Lindquist, Amsterdam: Elsevier, p. 347.

Jamiołkowski, A. (2000). *Rep. Math. Phys.* 46, 469.

Jamiołkowski, A. (2010), *Journal of Physics: Conference Series* 213, 012002.

Kossakowski, A. (1972) *Rep.Math. Phys.* 3, 247.

Kovačević, J. and Chebira, A. (2008). *An Introduction to Frames*, Boston-Delft, NOW.

Kraus, K. (1971). *Ann. Phys.* 64, 119.

Lindblad, G. (1976). *Comm. Math. Phys.* 48, 119.

Nielsen, M.A. & Chuang, I. (2000). *Quantum Computation and Quantum Information*, Cambridge, Cambridge Univ. Press.

Paris, M. & Rehacek J. (2004). *Quantum State Estimation*, Lect. Notes Phys. 649, Springer.

Shemesh, D. (1984). *Lin. Algebra Appl.* 62, 11.

Weigert, S. (2000). *New Insight in Quantum Mechanics*, Eds. H.-D. Doebner et al., Singapore, World Scientific.

Part 2

Quantum Phenomena with Laser Radiation

5

Quantum Optics Phenomena in Synthetic Opal Photonic Crystals

Vasilij Moiseyenko and Mykhailo Dergachov
Oles' Honchar Dnipropetrovsk National University
Ukraine

1. Introduction

Optical phenomena in materials with a space modulation of dielectric constant at distances close to the light wavelengths (so called photonic band-gap structures or photonic crystals) are of a great interest now because of the existence of band gaps in their photonic band structure (Bykov, 1972; Yablonovitch, 1987; John, 1987). These band gaps represent frequency regions where electromagnetic waves are forbidden, irrespective of the spatial propagation directions. Inside the band gaps, the photon density of states is equal to zero and so the emission of light sources embedded in these crystals should be inhibited in these spectral regions (Vats et al., 2002). Since the time the effect is predicted, many experiments have been devoted to studies of spontaneous emission of molecules embedded in photonic crystals (Gaponenko et al., 1999; Gorelik, 2007). Typical structures of photonic crystals and calculations of corresponding photonic band structures are presented in a book by Prof. Joannopoulos (Joannopoulos et al., 2008). Besides the emission inhibition effect, a number of new optical phenomena in 3D photonic crystals, interesting from the applied point of view, are under intensive study now. The main research directions are the following:

- Effects of light localisation (John, 1987; Kaliteevskii et al., 2005; Vignolini et al., 2008).
- Radiation of photonic crystals filled with organic (rhodamine 6G, 1,8-naphthoylene-1′,2′-benzimidazole, stilbene) and inorganic (ZnO, ZnS, rare-earth ions Eu^{3+}, Tb^{3+}, Er^{3+}) luminophores near by the edges of photonic band-gap (Gaponenko et al., 1999; Aliev et al., 2002; Emel'chenko et al., 2005; Li et al., 2007).
- Radiation of CdTe, CdSe/ZnS quantum dots in photonic crystal volume (Gruzintsev et al., 2009; Ambrozevich et al., 2009).
- Quantum optics phenomena in nano-structured materials based on photonic crystals and nonlinear optical substances (Lin & Vuckovic, 2010).
- Effects of the radiation field amplification in photonic crystals (Lin & Vuckovic, 2010).
- Increase of solar cells efficiency with the use of photonic crystals (Florescu et al., 2007).

As a good prototype of 3D photonic crystals, synthetic opals made of α-SiO_2 globules have been widely used (Gorelik, 2007). Diameter D of globules can be varied from 200 nm to 700 nm. Between globules there are tetrahedral and octahedral hollows (or pores) with a mean size of about $0.26D$. Synthetic opals are characterized by a stop-band or a pseudogap (i.e., a band gap actual for one direction) in the [111] direction and a singular behaviour of photon density of states near by the stop-band edges. The existence of pores in the opal structure allows modifying optical properties of such systems by filling the pores with various substances.

Synthetic opal photonic crystals containing nonlinear optical substances give a good chance to observe quantum optics phenomena in spatially nonuniform media where the photon mean free path is close to the light wavelength. Moreover, in this case the input optical power that is necessary to observe phenomena may be lower than the power required usually for observing the same phenomena in uniform nonlinear substances. The reason for it is the existence of diffuse transfer of photons that can result in photon accumulating inside photonic crystals and, consequently, in local optical power increasing. In particular, the possibility of experimental manifestation of Raman scattering and spontaneous parametric down-conversion in synthetic opals is discussed (Gorelik, 2007). The latter phenomenon is of special interest as it is convenient method to obtain bi-photon fields consisting of correlated photon pairs (Kitaeva & Penin, 2005). In the recent years, crystals with chirped structure of quadratic susceptibility (Kitaeva & Penin, 2004), and materials with spatially regular and stochastic distribution of quadratic optical susceptibility (Kalashnikov et al., 2009), are considered as sources of bi-photons. It is quite possible synthetic opal photonic crystals will be ranked with these sources.

2. Fundamentals of quantum optics phenomena in photonic crystals

Optical processes in nano-structured materials with a period close to the light wavelength are essentially different from those in bulk uniform media. It is due to the regularities of propagation of Bloch optical waves in such periodic structures ("photon confinement").

2.1 Luminescence

Consider the spontaneous emission transition in two-level system presented in Fig. 1.

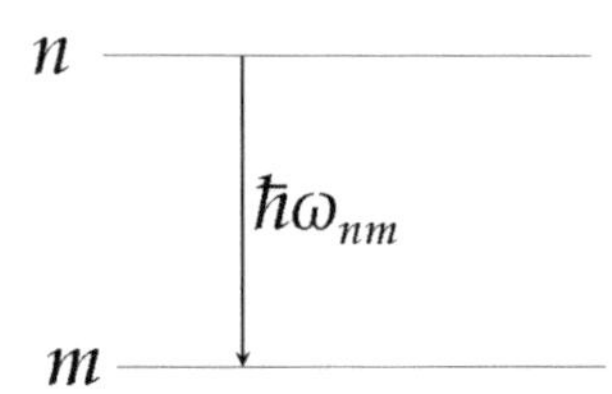

Fig. 1. Spontaneous emission transition in two-level system

The downward transition probability W_{nm} is determined in accordance with Fermi's golden rule as follows

$$W_{nm} = \frac{2\pi}{\hbar} \cdot \left| \langle m | \hat{H}_{\text{int}} | n \rangle \right|^2 \cdot g(E_n - \hbar\omega), \tag{1}$$

where $\left| \langle m | \hat{H}_{\text{int}} | n \rangle \right|$ is a matrix element of the perturbation operator, $g(E_n - \hbar\omega)$ is a density of final states of micro-object. In case of placing the micro-object into photonic crystal, the $g(E_n - \hbar\omega)$ spectrum is characterised by a density of optical states $g(\omega)$ in photonic crystal.

The spontaneous emission spectrum $S(\omega)$ is completely determined by the spectral distribution of transitions frequencies ω_{nm} and the density of optical states $g(\omega)$ within a region of these frequencies (Vats et al., 2002).

When frequencies ω_{nm} are in photonic band gap, where $g(\omega) = 0$, spontaneous emission must be completely inhibited. In general case, a dip in spontaneous emission spectrum should appear. The spectral position of this dip is correspondent to positions of reflection spectrum maximum and transmission spectrum minimum (Gaponenko et al., 1999). As a result of spontaneous emission inhibition the localisation of photon near by irradiative atom inside photonic crystal becomes possible if the transition frequency is within a band gap region or in the vicinity of band gap edges (John, 1987). In this case, bonded atom-photon state is coming into being. The photon emitted returns to the atom due to Bragg reflection and is re-absorbed by this atom. The existence of a group of such atoms may result in forming narrow photonic impurity band like an impurity band in semiconductor at sufficient concentrations of impurity atoms. Kinetics of luminescence in the vicinity of band gap edges demonstrates non-exponential behaviour (John, 1987).

2.2 Enhanced Raman scattering

Consider an elementary Stokes Raman scattering process as a disintegration of exciting photon ($\hbar\omega_{ex}$, $\boldsymbol{k}_{ex}$) into scattered photon ($\hbar\omega'$, $\boldsymbol{k}'$) and optical phonon ($\hbar\Omega$, $\boldsymbol{K}$) (Fig. 2).

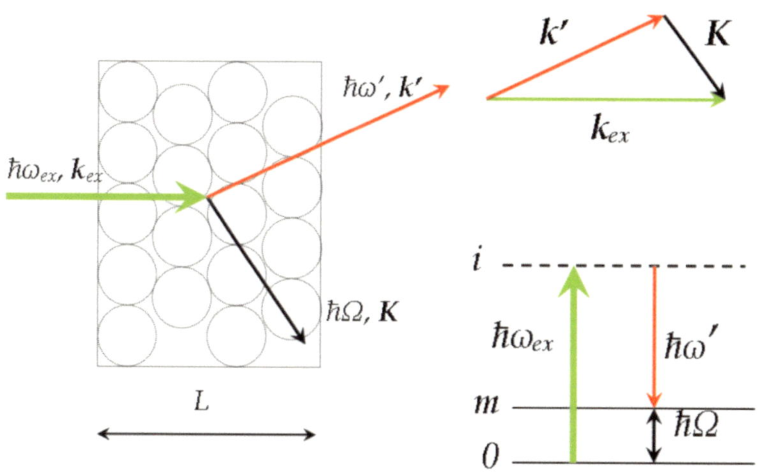

Fig. 2. Elementary Stokes Raman scattering process

The Stokes process probability $W(\omega')$ is determined by the density of optical states $g(\omega')$ in the region of scattered light frequencies (Poulet & Mathieu, 1970)

$$W(\omega') = \frac{2\pi}{\hbar} \cdot \left| \langle m | \hat{H}'_{\text{int}} | 0 \rangle \right|^2 \cdot g(\omega'), \tag{2}$$

where $\left| \langle m | \hat{H}'_{\text{int}} | 0 \rangle \right|$ is the modulus of the matrix element of the Hamiltonian of the radiation-substance interaction.

In general, the density of optical states $g(\omega)$ in photonic crystal is defined by a photon dispersion law and has maxima near by the band gap edges, where $(d\omega / dk) \rightarrow 0$. In the frame of one-dimensional model it can be determined as follows

$$g(\omega) = (k^2 / \pi^2)(d\omega / dk)^{-1}. \tag{3}$$

Exciting photons with the frequency ω_{ex} in the vicinity of band gap edges have the velocity values close to zero. It results in increasing the interval t'_{int} of radiation-substance interaction in photonic crystal compared with the interval t_{int} in uniform material, according to the following expression

$$t'_{\text{int}} = t_{\text{int}} \cdot \frac{c}{n_{eff}} \cdot (d\omega / dk)^{-1}, \tag{4}$$

where $c = 3 \cdot 10^8$ m/s and n_{eff} is an effective refractive index of photonic crystal. Besides, the interval t'_{int} may be enlarged because of the presence of "diffuse" photons whose motion is like to Brownian motion. Such diffuse photon transfer is most probably due to multiple photon reflections from disordered elements in photonic crystal structure. Both considered mechanisms give a reason to expect the enhancement of Raman scattering by substances infiltrated into photonic crystal.

2.3 Spontaneous parametric down-conversion

Spontaneous parametric down-conversion is a process of spontaneous disintegration of pump photons ($\hbar\omega_p$, $\boldsymbol{k}_p$) into pairs of signal ($\hbar\omega_s$, $\boldsymbol{k}_s$) and idler ($\hbar\omega_i$, $\boldsymbol{k}_i$) photons. As this process is a second-order nonlinear process it occurs in media with no inversion symmetry. In case of spatially uniform media with a non-zero second-order nonlinear susceptibility $\chi^{(2)}$ the energy and momentum conservation is as follows

$$\omega_p = \omega_s + \omega_i, \; \boldsymbol{k}_p = \boldsymbol{k}_s + \boldsymbol{k}_i \tag{5}$$

In frequency-degenerated ($\omega_s = \omega_i = \omega_p / 2$) and collinear ($k_s = k_i = k_p / 2$) regime at equal polarization of photons in pair (i.e., signal and idler photons are identical) the spectrum of bi-photons is determined by the following expression (Kalashnikov et al., 2009)

$$J(\Omega_{bp}) \sim \left[\frac{\sin x}{x}\right]^2, \tag{6}$$

where $x = (L/2) \cdot (d^2k / d\Omega_{bp}^2) \cdot \Omega_{bp}^2$, Ω_{bp} is a frequency turning out of $\omega_p / 2$, L is a sample length in the pump propagation direction.

For spatially nonuniform media with regular structures (photonic crystals) a periodic modulation of linear and nonlinear susceptibilities should be considered in general case. By taking into account the $\chi^{(2)}$ periodic modulation the bi-photons spectrum should be determined by an additive sum of single harmonics of $\chi^{(2)}$ susceptibility (Kitaeva & Penin, 2005). The spectrum of each of these harmonics is shifted relative to the spectrum in uniform medium, according to the following "quasi-synchronism" condition (see also Fig. 3)

$$\Delta \boldsymbol{k}_m \equiv \boldsymbol{k}_s + \boldsymbol{k}_i + m\boldsymbol{q} - \boldsymbol{k}_p = 0\,, \tag{7}$$

where m = 1, 2, 3 …, and $\boldsymbol{q}$ is a vector of photonic crystal reciprocal lattice ($q = 2\pi/d$, d is a period of photonic crystal structure).

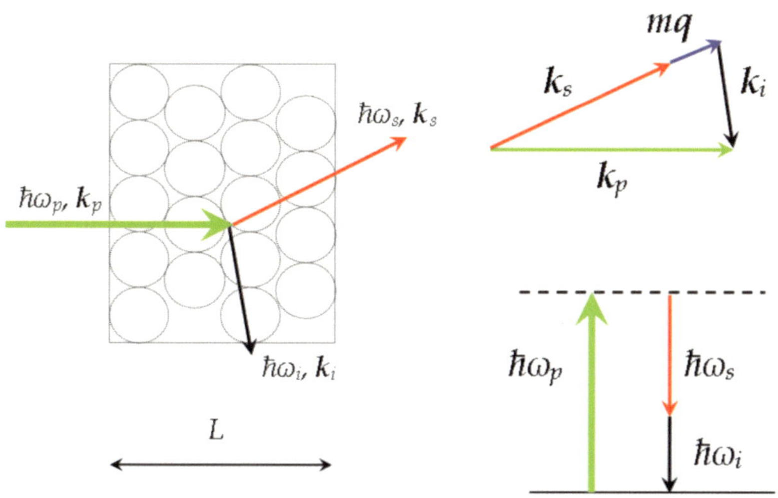

Fig. 3. Spontaneous parametric down-conversion process in photonic crystal

The intensity of each summand is determined by the square of magnitude of corresponding harmonic amplitude. An additional contribution to the parametric down-conversion spectrum should be given by interference of non-coinciding harmonics. At the absence of absorption the parametric down-conversion intensity $J_{\omega\Phi}$ per unit spectral $\Delta\omega$ and angular $\Delta\Phi$ intervals is determined by a magnitude of phase quasi-synchronism turning out Δ_m for the m-th order nonlinear diffraction (Kitaeva & Penin, 2004)

$$J_{\omega\Phi} \sim I_0 \cdot L^2 \sum_m \left| \chi_m^{(2)} \frac{\sin(\Delta_m / 2)}{(\Delta_m / 2)} \right|^2 \tag{8}$$

where I_0 is a pump intensity, $\chi_m^{(2)}$ is an amplitude of the m-th Fourier component of spatial distribution of nonlinear susceptibility $\chi^{(2)}$. The phase quasi-synchronism turning out Δ_m is defined via the phase synchronism turning out Δ in spatially uniform medium as follows

$$\Delta_m = \Delta - mqL\,. \tag{9}$$

In contrast to parametric down-conversion spectrum of spatially uniform sample the bi-photon field spectrum of photonic crystal should be broadened, and the interference effects may appear in its spectral intensity distribution (Kitaeva & Penin, 2004, 2005; Nasr et al., 2008; Kalashnikov et al., 2009). One of the reasons that cause an additional broadening of the bi-photon spectrum in photonic crystal is the presence of structure disordered domains with period d varied along a pump propagation direction. Besides, observable spectrum should be determined by the density of optical states $g(\omega)$ in the region of scattered light frequencies as a result of periodical modulation of linear susceptibility.

3. Characterisation of samples and experimental setup

Nanodisperse globules of silica dioxide were synthesized using a modified Stöber method (Stöber et al., 1968) through hydrolysis of tetraethoxysilane $Si(OC_2H_5)_4$ at high values of water concentration. Bulk synthetic opals were obtained by natural sedimentation of α-SiO_2 globules with the following annealing of samples at 800 ^{0}C during several hours. Annealing was performed in order to remove organic residua, extra- and intra-globular ethoxygroups, and chemically bound water (Samarov et al., 2006). Dimensions of obtained samples were about 1.0x1.0x0.2 cm^3.

Characterization of initial opals was performed by analyzing the surface structure with the use of X-Ray Microanalyzer JEO JXA 8200 and by measuring transmission and reflection spectra within a visible spectral range (Fig. 4, 5). Opal samples in these studies were composed of hexagonal close-packed layers of monodisperse α-SiO_2 globules which are arranged in the face centred cubic lattice. The value of globules diameter D in various samples was from 250 nm to 270 nm, and the distance d between the (111) planes was from 204 nm to 220 nm.

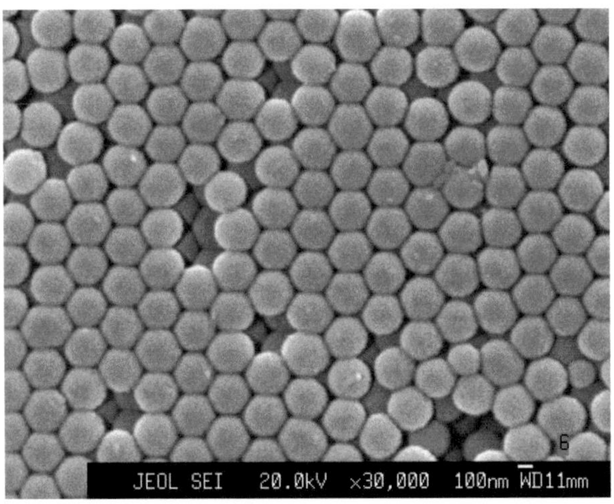

Fig. 4. Image of the opal surface in [111] direction. Diameter of globules D = 270 nm

Parameters of photonic stop-band in [111] direction (spectral position of stop-band centre λ_c and its spectral width $\Delta\lambda_g$) were determined from transmission and reflection spectra as the parameters of non-transmission or reflection band (spectral position of maximum and spectral band width) formed in accordance with Bragg diffraction mechanism. The spectral position of reflection maximum (or transmission minimum) assigned as stop-band centre λ_c is dependent of an incident angle θ, effective refractive index n_{eff} and distance d between the (111) planes as follows (Podolskyy et al., 2006):

$$\lambda_c(\theta) = 2d\sqrt{n_{eff}^2 - \sin^2\theta}\,. \tag{10}$$

The effective refractive index n_{eff} is determined by the refractive index n_s of α-SiO_2 globules (n_s = 1.47), the refractive index n_p of substance in opal pores (for initial opal it is air, n_p = 1) and the volume fraction f occupied by α-SiO_2 globules (in our case, $f \approx 0.74$) as follows:

$$n_{eff}^2 = f \cdot n_s^2 + (1-f) \cdot n_p^2 . \tag{11}$$

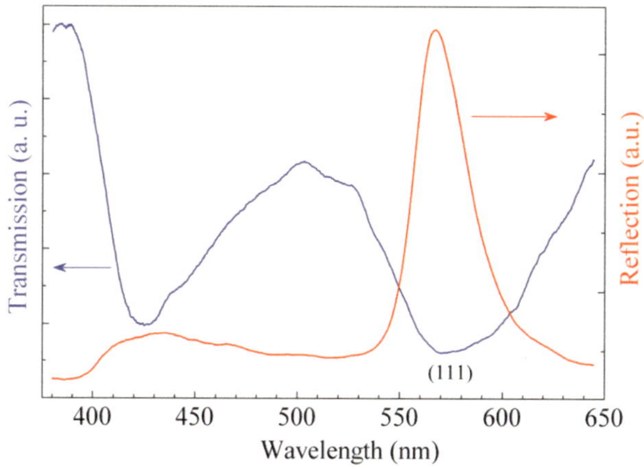

Fig. 5. Typical transmission and reflection spectra of initial opals (D = 255 nm). Transmission spectrum was measured at normal incidence to (111) plane ($\theta = 0^0$), reflection spectrum was measured at $\theta = 7^0$

Photonic crystals based on synthetic opals were obtained by further infiltration of initial opals with organic luminophores (rhodamine 6G, 2,5-bis(2-benzoxazolyl)hydroquinone, pironin G, astrofloksin) or nonlinear optical substances ($Ba(NO_3)_2$, $LiIO_3$, KH_2PO_4, $Li_2B_4O_7$). In most cases the infiltration was performed by a multiple soaking of samples in corresponding supersaturated solutions at room temperature. For example, synthetic opals were filled with rhodamine 6G by soaking samples in a dilute ethanol solution with laser dye concentrations of 10^{-4} M or $5 \cdot 10^{-3}$ M. After soaking the obtained samples were in the air until ethanol was evaporated. In case of infiltration with $Ba(NO_3)_2$, $LiIO_3$, KH_2PO_4 an additional annealing of samples was performed at temperatures lower than melting ones (595 ^{o}C for $Ba(N0_3)_2$ and 120 ^{o}C for $LiIO_3$) to remove water. In case of $Li_2B_4O_7$ the initial opal was in $Li_2B_4O_7$ melt at 860 ^{0}C.

Reflection and transmission spectra of opals after infiltration were measured to prove the existence of corresponding substance in pores. Two types of changes in the spectra were registered. First, in opals with organic luminophores, an additional non-transmission band caused by absorption of embedded molecules was observed (Fig. 6). Second, the band caused by Bragg diffraction was shifted if a quantity of embedded substance was enough to change essentially the value of n_{eff}, according to expression (11) (Podolskyy et al., 2006).

In some experiments, in order to diminish (or exclude) the photonic stop-band effects and to study phenomena in a regular matrix of nano-emitters the opal samples were additionally soaked in water-glycerine solutions or pure glycerine. Opal infiltration with any water-glycerine solution yields in decreasing dielectric contrast in the synthetic opal photonic crystals as the refractive index of a water-glycerin solution n_p (variable from 1.39227 till 1.47399 in our experiments) is close to that of SiO_2 globules n_S. It causes the shift of the stop-band center λ_c to the longer wavelengths (10) and the narrowing of stop-band region $\Delta\lambda_g$ with increasing glycerin concentration.

For studying luminescence and light scattering phenomena in synthetic opals photonic crystals the incoherent and coherent light sources were used. The incoherent light sources were two light emitting diodes Edixeon EDST-3LAx (λ = 400 nm and 517 nm, $\Delta\lambda_{1/2} \approx$ 30 nm, and the average power P = 30 mW). The coherent light sources were the pulsed nitrogen laser (λ = 337 nm, $\Delta\lambda_{1/2}$ = 0.1 nm) with the pulse repetition frequency of 100 Hz and P = 3 mW, the semiconductor laser (λ = 407 nm, $\Delta\lambda_{1/2}$ = 1 nm, P = 60 mW), and the diode-pumped solid state laser (λ = 532 nm, $\Delta\lambda_{1/2}$ = 1 nm, P = 120 mW). As a rule, the forward and back scattering geometry along the [111] direction were used. Length of samples along the excitation direction was from 2 mm to 3 mm. Some experiments on light scattering were performed in the right angle geometry. The secondary emission from the sample surface was collected along the [111] direction by using lens with an aperture of about 0.17·π sr. The angular dependences of emission spectra were obtained within an angles region from 1^0 to 5^0 with the use of circle diaphragms. Spectral analysis was performed by using modernized spectrometer DFS-12. Signal registration was carried out in a regime of photon counting with accumulation.

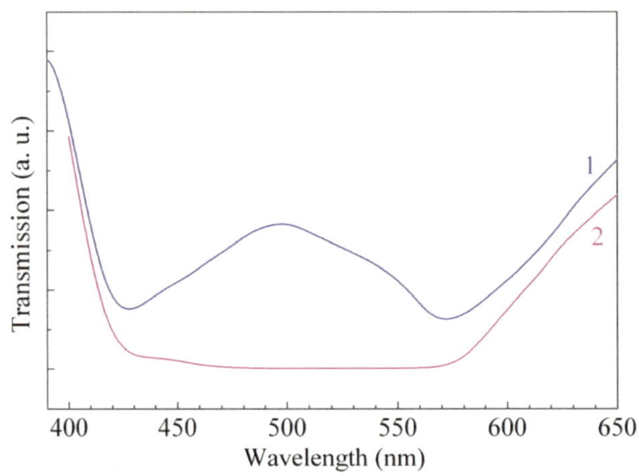

Fig. 6. Transmission spectra of initial opal (1) and the same opal infiltrated with pironin G (2). Spectra were measured at normal incidence to (111) plane ($\theta = 0^0$)

4. Results and discussion

In emission spectra of synthetic opal photonic crystals under optical excitation three typical regions are clearly observed (Gaponenko et al., 1999; Gorelik, 2007; Gruzintsev et al., 2008). One of them takes place in the opal-luminophores spectra, and is beyond doubt caused by the irradiative transitions between luminophore molecule levels. The other regions are inherent in emission spectra of opals filled with nonlinear optical substances. The first one is in a spectral range typical for Raman scattering region. The position of the second one is more distanced from an exciting line and is rather correlated with a stop-band position.

4.1 Luminescence of organic molecules in synthetic opals

As mentioned above synthetic opals are characterised by a presence of band gap in one space direction. This is why a complete inhibition of spontaneous emission should be absent. Nevertheless, the use of opals as containers of emitting organic molecules allows changing the irradiative transitions probabilities remarkably (Bechger et al., 2005). It may be actual for substances with intra-molecular proton transfer, such as 2,5-bis(2-benzoxazolyl)hydroquinone, to control the probabilities of transitions with and without proton transfer.

4.1.1 Laser dyes molecules

Luminescence spectra of laser dyes (rhodamine 6G and pironin G) are shown in Fig. 7 (Moiseyenko et al., 2008, 2010). As seen from Fig. 7, for both molecules embedded into opal matrix a partial inhibition of luminescence intensity takes place within a region corresponding to the stop-band. For the rhodamine 6G spectrum, enhancement of the short-wavelength tail of the emission band is observed, while the long-wavelength tail of the spectrum is not altered essentially (Fig. 7, a). At the same time, the opal-pironin G spectrum is concentrated inside a long-wavelength region though it is rather not amplified (Fig. 7, b).

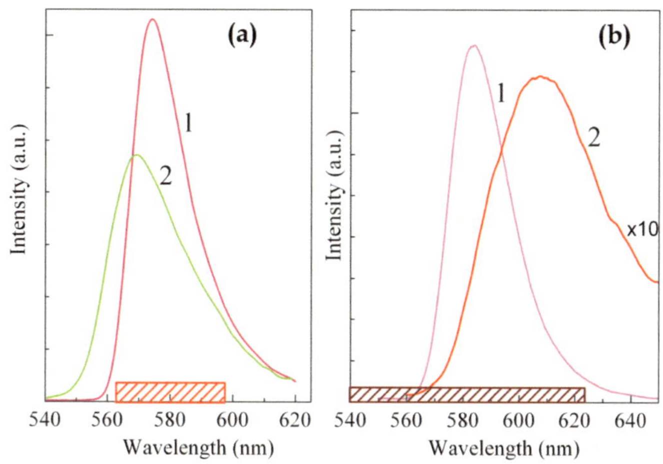

Fig. 7. Luminescence spectra of rhodamine 6G (a) and pironin G (b) put into optical cell with an ethanol (1) and infiltrated into opals (2) at a 517 nm diode excitation. Rectangles point to stop-band positions.

Following Bechger et al., 2005, observed transformations may be explained in such a way. When the light with wavelength λ_{em} shorter than a stop-band centre λ_c is emitted in the [111] direction it encounters Bragg diffraction at higher angles. Because of diffuse propagation more light is detected in the [111] direction at these higher angles. For the light of any longer wavelengths ($\lambda_{em} > \lambda_c$) all the directions in opal volume are equivalent and that light escapes the sample without being enhanced. As mentioned above, spectral distribution of spontaneous emission is defined by density of photon states $g(\omega)$. In both cases the spectral intensity maximum is near by the stop-band edge where the density of states $g(\omega)$ has maximum (Fig. 7). An absence of total inhibition inside stop-band region can be connected with a structure disorder that results in appearance of local states in the stop-band (Kaliteevskii et al., 2005). From these points of view luminescence spectra of rhodamine 6G and pironin G in opals with additional water-glycerine solutions (Fig. 8) may be interpreted as follows. Infiltration with any water-glycerine solution results in lowering dielectric contrast. It causes a shift of stop-band centre λ_c to longer wavelengths. In accordance with our calculations by using expressions (10, 11), the stop-band shift is equal to 6 nm by varying glycerine volume concentration from 66 % till 100 %. It corresponds exactly to the luminescence maximum shift observed in the opal-rhodamine 6G spectrum (Fig. 8, a). In case of pironin G, we have somewhat different behaviour (Fig. 8, b, and Fig. 9).

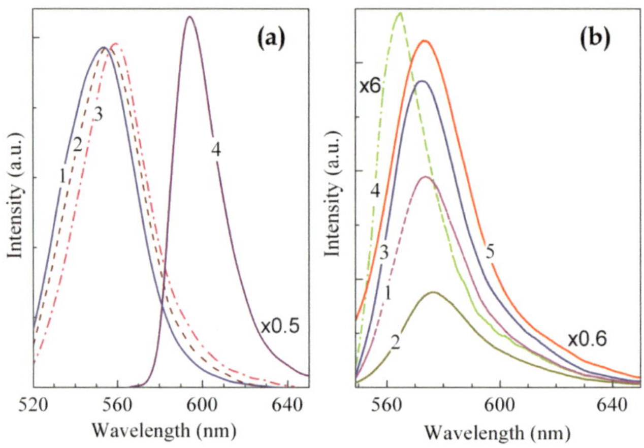

Fig. 8. Luminescence spectra of rhodamine 6G (a) and pironin G (b) placed into opals filled with a water-glycerin solution and in the optical cell with pure glycerine. In case (a) glycerine volume concentrations are 66 % (1), 75 % (2), 100 % (3), and curve 4 is the spectrum in pure glycerine. In case (b) glycerine volume concentrations are a 40% (1), 60% (2), 80% (3), 100% (4), and curve 5 is the spectrum in pure glycerine.

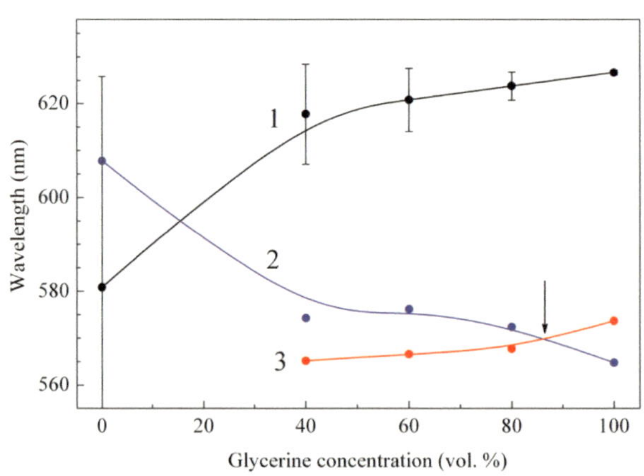

Fig. 9. Concentration dependences of stop-band center λ_c (1), pironin G luminescent maxima in opals (2) and water-glycerine solution in optical cell (3). The bars are stop-band widths.

Without infiltrating opals with a solution the condition $\lambda_{em} > \lambda_c$ takes place and we have a weak emission in the long-wavelength region discussed above (Fig. 7, b). By increasing glycerine concentration the relation between λ_{em} and λ_c becomes the opposite ($\lambda_{em} < \lambda_c$) and the short-wavelength tail of the luminescence band is being enhanced. At glycerine concentrations close to 85 volume per cents (marked by the arrow in Fig. 9), the dielectric contrast vanishes ($\Delta\lambda_g = 0$) and luminescent band position becomes just the same as in the optical cell (a so-called solvent effect). After passing through this concentration, the stop-band width $\Delta\lambda_g$ starts growing due to increasing refractive index n_p. In this case we have an inversion of photonic bands, something like that occurring in narrow-gap semiconductors. The luminescent band in such inverse opal is shifting towards a "blue" side.

4.1.2 Intra-molecular proton transfer substances

2,5-bis(2-benzoxazolyl)hydroquinone belongs to a class of substances that manifests intra-molecular excited-state proton transfer. This substance is tautomerized in the conditions of ultraviolet excitation and shows a pronounced luminescence in green-red region with a large Stokes shift. When 2,5-bis(2-benzoxazolyl)hydroquinone in a hexane solution is excited within a main absorption band (280 nm - 420 nm), the irradiative transitions in both structural forms appear the spectrum (curve 1 in Fig. 10). The band in the 430 nm - 470 nm region is correspondent to the transitions without proton transfer, the band in the 580 nm - 620 nm region is due to the transitions with proton transfer. In condensed states these bands may be shifted towards the greater wavelength region. Thus a wide intensive band observed in the polycrystalline state spectrum within a 600 nm - 750 nm region is a result of the shift of a "proton-transfer" band. A shoulder of this band (in the 490 nm - 560 nm region) is most likely due to the impurity luminescence. It is proved by diminishing this band intensity in amorphous state (impurity-free state, according to our obtaining procedure) and by results presented by Chayka et al., 2005.

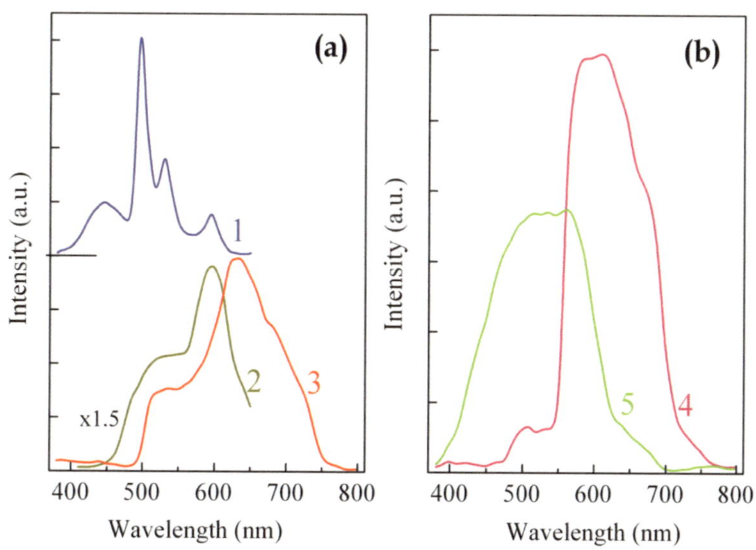

Fig. 10. Luminescence spectra of 2,5-bis(2-benzoxazolyl)hydroquinone in a hexane solution at a 400 nm diode excitation (1), in polycrystalline state at a 350 nm wide-band mercury lamp excitation (2), in polycrystalline (3) and amorphous (4) states, and into synthetic opal volume (5) at a 337 nm nitrogen laser excitation.

Spectral intensity distribution in the spectrum of 2,5-bis(2-benzoxazolyl)hydroquinone in synthetic opal is like to that in amorphous state (curves 4, 5 in Fig. 10). It allows assuming amorphous state of the substance in opal pores. The "blue" shift observed in this case may be explained in the following way. As a "proton-transfer" band is near by the stop-band region (600 nm - 640 nm in opal under study), the probability of these transitions decreases. It may result in increasing probabilities of impurity irradiative transitions and transitions without proton transfer. The latter transitions have not been observed in a "free" condensed state (Chayka et al., 2005). Another reason to make these processes observable is an accumulation of the shorter wavelength radiation because of Bragg reflection from the {111} planes at higher incident angles (Bechger et al., 2005).

4.2 Light scattering by synthetic opals filled with nonlinear optical dielectrics

Emission spectrum of synthetic opals with dielectrics observed within a wide spectral region under ultraviolet excitation may be divided into two parts. The first part is the spectrum located in the vicinity of the excitation line. The second one is in a region of 440 nm - 650 nm including a stop-band. By now the secondary emission of initial and dielectrics infiltrated opals has been given no universal explanation which should be satisfied by all experimental facts (Gorelik, 2007; Gruzintsev et al., 2008; Moiseyenko et al., 2009a, 2009b, 2012).

4.2.1 Enhanced spontaneous Raman scattering

Consider a total emission spectrum of initial synthetic opals; some of them were in air for a long time. The others were excited just after high temperature annealing. In all spectra a quite intensive band in the vicinity of the excitation line was observed. Its spectral position was independent of the stop-band position and previous technology conditions (Fig. 11). In the spectrum of a long time air-conserving sample a wide band with maximum at 570 nm was also observed. The fact that the band has vanished after annealing, i.e. after removing water in opal pores, reveals the impurity OH-groups luminescence origin of this band.

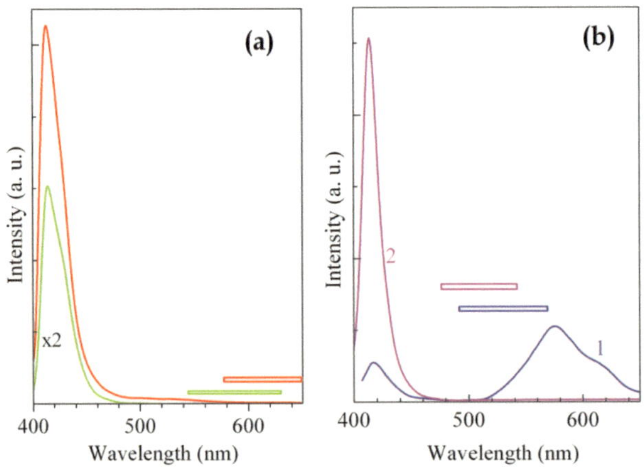

Fig. 11. Emission spectra of two different initial opals after thermal annealing at 800 ^{0}C (a). Emission spectra of opal conserved for a long time in the air at 70 % moisture (b) before (1) and just after (2) annealing at 700 ^{0}C. The rectangles point to the stop-band positions.

In order to understand the nature of the band near by the excitation line an influence of infiltrated substance on the emission spectrum has been studied (Moiseyenko et al., 2009a, 2009b). These spectra measured under spectral correct conditions with a 2 cm^{-1} resolution are presented in Raman shift scale after subtracting excitation line profile (Fig. 12, a). As seen from Fig. 12, spectral intensity distribution is dependent of kind of substance into opal pores. This fact together with mentioned above regularities allows us to suppose that the band observed within a typical vibrational spectrum range is caused by Raman scattering in substances forming photonic crystal. Such process becomes possible to be detected owing to an essential increase of field due to a slow diffuse motion of exciting photons into opal volume, and also, as a result of surface enhanced conditions inside opal pores.

However, obtained spectra are too wide compared with the usual Raman spectra. It may be explained, if remember, that band spectral profile is determined by spectral profile of excitation line and Raman spectrum of substance. In our initial experiments we have used a source with a significant width of the exciting line ($\Delta\lambda_{1/2} \approx 30$ nm). Another reason for spectrum broadening is a possible amorphous state of substances which form the sample structure. In case of amorphous state, a density of vibrational states $g(\Omega)$ can be quantitatively described by calculating reduced Raman spectrum $J_R(\Omega)$ for the Stokes component (Cardona, 1975) (Fig. 12, b).

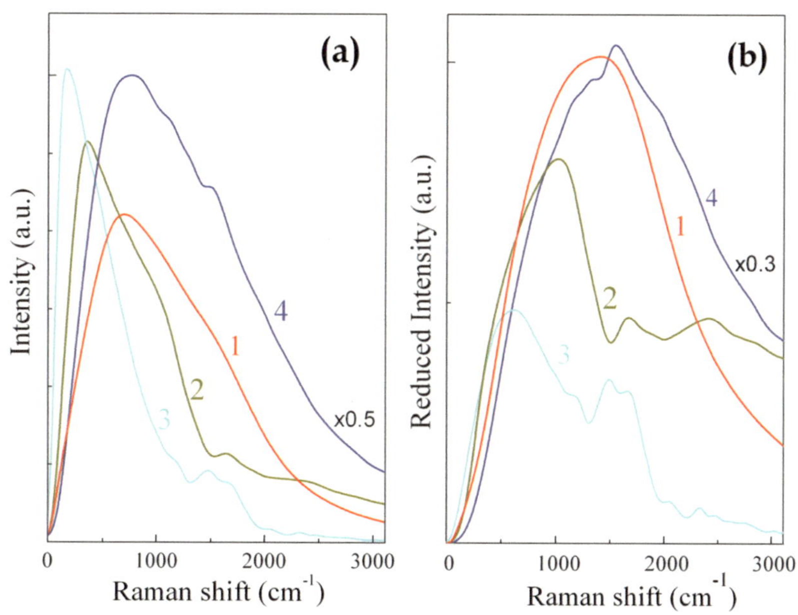

Fig. 12. Emission spectra (a) in the vicinity of the 400 nm exciting line and the corresponding reduced Raman spectra (b) for initial synthetic opal (1) and opals infiltrated with $CuCl_2$ (2), $Ba(NO_3)_2$ (3), and $LiIO_3$ (4)

To diminish a role of exciting radiation parameters in forming measured spectrum, we have used a 532 nm laser radiation with $\Delta\lambda_{1/2} \approx 1$ nm to excite emission in opal filled with KH_2PO_4 (Fig. 13). The significant band width in this case may testify amorphous state of substance in opal pores. The presence of the anti-Stokes component should be pointed out.

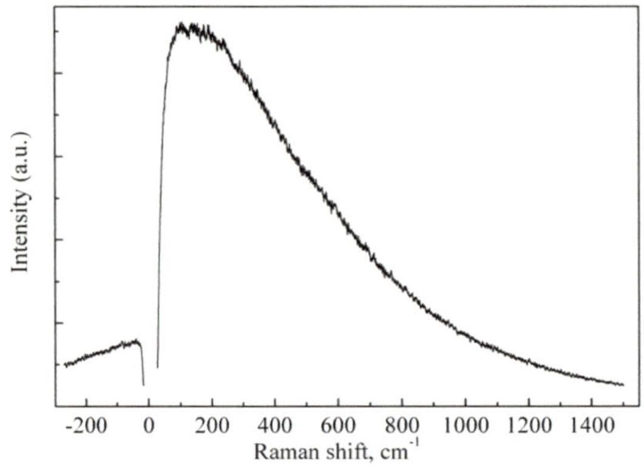

Fig. 13. Emission spectrum in the vicinity of the 532 nm laser exciting line for synthetic opal infiltrated with KH_2PO_4

The emission spectral distribution typical for Raman spectrum was observed in the opal-$Li_2B_4O_7$ spectrum in the right angle geometry (Fig. 14). The A_1(TO) Raman $Li_2B_4O_7$ spectrum measured earlier (Moiseyenko et al., 2006) at the excitation of a 532 nm Q-switched Nd:YAG laser with mean power of 250 mW is also presented in Fig. 14.

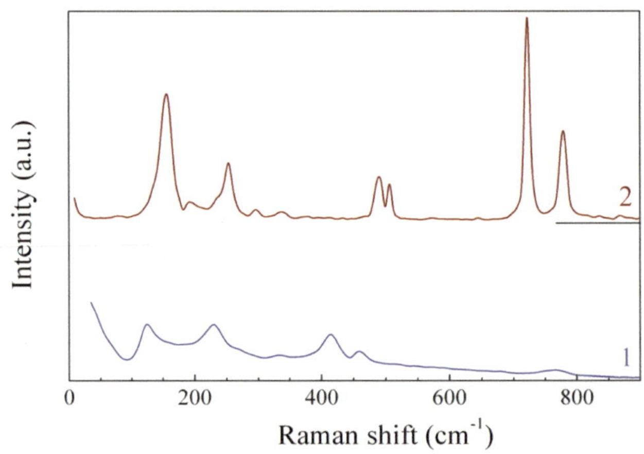

Fig. 14. The emission spectrum of opal infiltrated with $Li_2B_4O_7$ (1) and A_1(TO) Raman $Li_2B_4O_7$ spectrum (2) at the 532 nm laser excitation

Both spectra have a similar structure in the 100 cm^{-1} - 550 cm^{-1} spectral range, but the bands in the opal-$Li_2B_4O_7$ spectrum are shifted towards the excitation line and have a greater halfwidth. However, the values of bands halfwidths (no more than 30 cm^{-1}) give no reason to conclude amorphous state of the substance in opal pores. The broadening of bands is rather caused by structural disordering and the existence of polydomain structure. The bands shifts are most probably due to the small sizes of the unit $Li_2B_4O_7$ scattering volume defined by the pores sizes (no more than 100 nm in our samples). The coincidence of high-frequency Raman range and the stop-band spectral region results in a crucial decrease of emission intensity at Raman shifts higher than 600 cm^{-1}.

In order to experimentally prove the enhancement effects in synthetic opal photonic crystals Raman spectra in opal-$Li_2B_4O_7$ and single $Li_2B_4O_7$ crystal were measured in the low-frequency region under the same conditions (Fig. 15). As seen from Fig. 15, integral scattering intensity in the opal-$Li_2B_4O_7$ spectrum is about of a three times higher than the one in the single $Li_2B_4O_7$ crystal spectrum. Taking into account the lesser quantity of lithium tetraborate in opal matrix in the same scattering volume (no more than 26 % from total volume, as lithium tetraborate is situated only in opal pores) we can estimate the Raman enhancement coefficient as high as 10. Two enhancement mechanisms can be proposed. The first one is a photon slowing in accordance with a dispersion law in photonic crystals and the second is a multiple reflection from disordered planes resulting to diffuse photon motion (Gorelik, 2007).

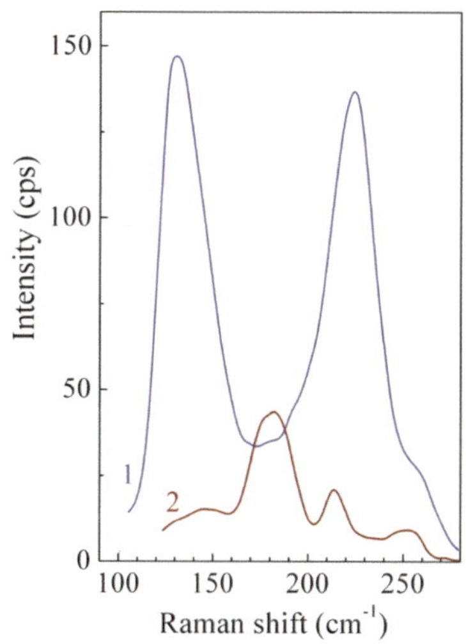

Fig. 15. Low-frequency region in non-polarised Raman spectra of opal infiltrated with $Li_2B_4O_7$ (1) and single $Li_2B_4O_7$ crystal (2) under the same conditions at the 532 nm diode pumped solid state laser excitation. The right angle geometry was used. The longer wavelength tale of excitation line was subtracted.

4.2.2 Spontaneous parametric down-conversion

Emission spectra of photonic crystals infiltrated with any nonlinear optical substances mentioned above are similar after subtracting the longer wavelength tale of excitation line and the bands corresponding to Raman scattering processes (Fig. 16). The spectra contain a wide asymmetric band within a 410 – 600 nm range. This band spectral position is different for opals with different infiltrators but it is correlated with the stop-band position. The emission intensity decreases within a stop-band region but it does not vanish completely because of the existence of point defects and structural disordering in photonic crystals.

Photos of secondary emission made far from sample surface reveal the angular distribution of spectral intensity (Moiseyenko et al., 2009b). Emission spectra measured at different scattering angles and treated by subtracting the longer wavelength tale of excitation line and

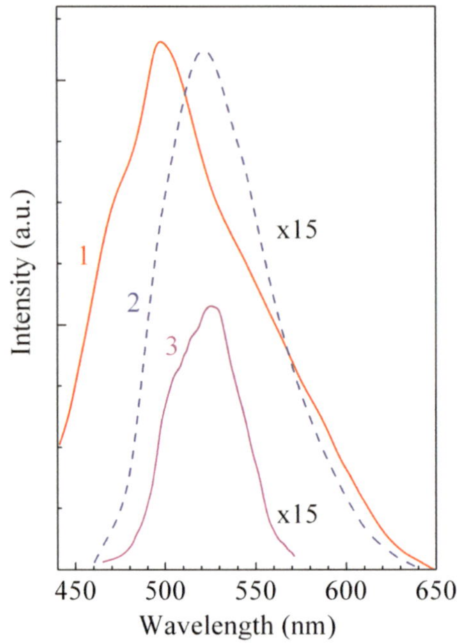

Fig. 16. Emission spectra of opals infiltrated with $Ba(NO_3)_2$ (1), $LiIO_3$ (2), and KH_2PO_4 (3) at a 400 nm diode excitation in the forward scattering geometry. Spectral stop-band positions are 525 nm - 590 nm (1), 560 nm - 615 nm (2, 3).

bands in Raman scattering region are presented in Fig. 17 (Moiseyenko et al., 2012). The spectral intensity distribution and the emission maximum position are quite different at various scattering angles. When dielectric contrast becomes negligible (by infiltrating opal with pure glycerine) a shifted symmetric emission band is observed (curve 4 in Fig. 17). In this case the sample may be considered as practically transparent matrix with periodic distributed nonlinear substance which is responsible for generating secondary emission. The symmetric form of spectral distribution is typical for spontaneous parametric down-conversion in uniform media (Kitaeva & Penin, 2005).

Thus, all elicited regularities together with chosen conditions of the samples heat treatment (Samarov et al., 2006) are the reasons to exclude the fluorescence of nano-composite components and the OH-groups fluorescence observed in synthetic opals within a 520 nm - 650 nm range (Gruzintsev et al., 2008). Taking into account angular dependences of spectral intensity, the emission observed may be interpreted as spontaneous parametric down-conversion in spatially nonuniform nonlinear optical media. Additional contribution to the emission within a 407 - 437 nm region may be given by enhanced Raman scattering discussed above.

As shown earlier, spontaneous parametric down-conversion intensity per a unit angle and spectral interval is determined by the value of quasi-synchronism Δ_m for the m-th order nonlinear diffraction. In case of 3D photonic crystals based on synthetic opals this magnitude is defined by the structure disordering degree, the nonlinear substance filling

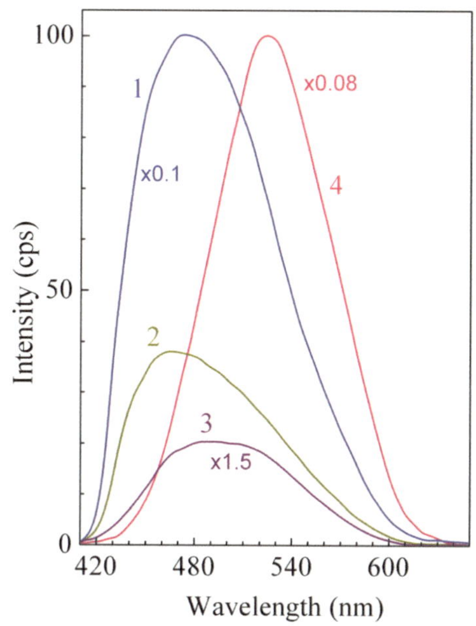

Fig. 17. Emission spectra of opal infiltrated with $Ba(NO_3)_2$ at a 407 nm laser excitation in the forward scattering geometry. The spectrum (1) was measured within a full angle range. The spectra (2) and (3) were registered at the angles of 5^0 and 3^0 to the pump propagation direction, respectively. The spectrum (4) was measured within a full angle range after additional sample infiltration with glycerine.

factor and by the existence of polydomain structure which forms additional superlattice. In our samples typical domain size was about of 70 mkm. Then the phase quasi-synchronism condition becomes true for the greater number of directions and wavelengths. It results in broadening the parametric down-conversion spectrum like that occurred in chirped structures with quadratic nonlinearity (Nasr et al., 2008). Besides spectrum broadening effects, nonlinear diffraction by 3D grating of quadratic optical susceptibility and bi-photon field interference may result in more complicated changes in spatial and frequency distribution of far field compared with that observed in polydomain crystals (Kitaeva & Penin, 2004).

5. Conclusion

In accordance with Fermi's golden rule, modification of luminescence spectra of organic molecules in synthetic opals is determined by the density of optical states in the vicinity of photonic stop-band. Inside stop-band region the partial inhibition of spontaneous emission is observed for all substances. If the wavelength, corresponding to the intensity maximum in spectrum of "free-state" substance, is shorter than wavelength, corresponding to the stop-band center, the amplification of dye luminescence at the high-energy edge of stop-band occurs. In the contrary case a weak luminescence is observed at the low-energy edge of stop-band without any amplifying.

The possibility of practical controlling probabilities of irradiative transitions without and with proton transfer by choosing properly the stop-band spectral position has been demonstrated. The "blue" shift of the luminescence spectrum of rhodamine 6G has been observed in opal additionally infiltrated with pure glycerine.

The emission band observed near by the exciting line is most probably due to Raman scattering in substances forming photonic crystal structure. It becomes possible to be detected owing to the essential increase of radiation field caused by the slow diffuse transfer of pump photons into sample volume.

Secondary emission of nonlinear photonic crystals under coherent and incoherent optical pumping observed within a 410 – 600 nm range has an asymmetric continual spectrum with a width of about 200 nm. The spectral intensity distribution and the intensity maximum position are dependent of the exciting radiation parameters (wavelength of excitation, degree of coherence, angle range of pumping wave vectors), of the emission detection angle, of the structure disordering degree. The emission observed is analyzed in terms of spontaneous parametric down-conversion phenomenon which occurs in spatially nonuniform nonlinear medium.

6. Acknowledgment

This work was financially supported by National Academy of Sciences of Ukraine, the Ukrainian-Russian project No.71-02-10 "Radiation of 3D photonic crystals under optical and electric excitations".

7. References

Aliev, G., Golubev, V., Dukin, A., Kurdyukov, D., Medvedev, A., Pevtsov, A., Sorokin, L., Hutchison, J. (2002). Structural, photonic band-gap, and luminescence properties of the opal-erbium composite. *Physics of the Solid State*, Vol.44, No12 (December 2002), pp. 2224-2231, ISSN 1063-7834

Ambrozevich, S., Gorelik, V., Dirin, D., Vasil'ev, R., Vitukhnovsky, A., Voinov, Yu. (2009). Optical properties of 3D photonic crystals filled with CdSe/CdS quantum dots. *Journal of Russian Laser Research*, Vol.30, No4, (July 2009), pp. 384-391, ISSN 1071-2836

Bechger, L., Lodahl, P., Vos, W. (2005). Directional fluorescence spectra of laser dye in opal and inverse opal photonic crystals. *Journal of Physical Chemistry B*, Vol.109, No5, (May 2005), pp. 9980-9988

Bykov, V. (1972). Spontaneous emission in a periodic structure. *Soviet Physics – JETP (Journal of Theoretical and Experimental Physics)*, Vol.35, No.2, (February 1972), pp. 269-273, ISSN 0038-5646

Cardona, M. (1975). *Light Scattering in Solids*, Springer-Verlag, Berlin

Chayka, K., Moiseyenko, V., Mordzinski, A. (2005). Luminescence of 2,5-bis(2-benzoxazolyl)hydroquinone molecules adsorbed on copper island film. *Ukrainian Journal of Physical Optics*, Vol.5, No4 (November 2005), pp. 128-132, ISSN 1609-1833

Emel'chenko, G., Gruzintsev, A., Koval'chuk, M., Masalov, V., Samarov, E., Yakimov, E., Barthou, C., Zver'kova, I. (2005). Opal – ZnO nanocomposites: Structure and emission properties. *Semiconductors*, Vol.39, No11, (November 2005), pp. 1328-1332, ISSN 1063-7826

Florescu, M., Lee, H., Puscausu, I., Pralle, M., Florescu, L., Ting, D., Dowling, J.P. (2007). Improving solar cell efficiency using photonic band-gap materials. *Solar energy Materials & Solar Cells*, Vol.91, (June 2007), pp. 1599-1610, ISSN 0927-0248

Gaponenko, S., Bogomolov, V., Petrov, E., Kapitonov, A., Yarotsky, D., Kalosha, I., Eychmuller, A., Rogach, A., McGlip, J., Woggon, U., Gindele, F. (1999). Spontaneous emission of dye molecules, semiconductor nanocrystals, and rare-earth ions in opal-based photonic crystals. *Journal of Lightwave Technology*, Vol.17, No.11, (November 1999), pp. 2128-2137, ISSN 0733-8724

Gorelik, V., (2007). Optics of globular photonic crystals. *Quantum Electronics*, Vol.37, No5, (May 2007), pp. 409-432, ISSN 1063-7818

Gruzintsev, A., Emel'chenko, G., Masalov, V., Romanelli, M., Barthou, C., Benalloul, P., Maitre, A. (2008). Luminescent properties of synthetic opal. *Inorganic materials*, Vol.44, No2, (February 2008), pp. 159-164, ISSN 0020-1685

Gruzintsev, A., Emel'chenko, G., Masalov, V., Yakimov, E., Barthou, C., Maitre, A. (2009). Luminescence of CdSe/ZnS quantum dots infiltrated into an opal matrix. *Semiconductors*, Vol.43, No2, (February 2009), pp. 197-201, ISSN 1063-7826

Joannopoulos, J., Johnson, S., Winn, J., Meade, R. (2008). *Photonic Crystals: Molding the Flow of Light* (second edition), Princeton University Press, ISBN 978-0-691-12456-8, Princeton and Oxford

John, S. (1987). Strong localization of photons in certain disordered dielectric superlattices. *Physical Review Letters*, Vol.58, No23, (June 1987), pp. 2486-2489, ISSN 0031-9007

Kalashnikov, D., Katamadze, K., Kulik, S. (2009). Controlling the spectrum of a two-photon field: Inhomogeneous broadening due to a temperature gradient. *JETP Letters*, Vol.89, No5, (May 2009), pp. 224-228, ISSN 0021-3640

Kaliteevskii, M., Nikolayev, V., Abram, R., (2005). Eigenstate statistics and optical properties of one-dimensional disordered photonic crystals. *Physics of the Solid State*, Vol.47, No10, (October 2005), pp. 1948-1957, ISSN 1063-7834

Kitaeva, G., Penin, A. (2004). Diagnostics of the inhomogeneous distribution of quadratic optical susceptibility over parametric scattering spectra. *Quantum Electronics*, Vol.34, No7, (July 2004), pp. 597-611, ISSN 1063-7818

Kitaeva, G., Penin, A. (2005). Spontaneous parametric down-conversion. *JETP Letters*, Vol.82, No6, (September 2005), pp. 350-355, ISSN 0021-3640

Li, M., Zhang, P., Li, J., Zhou, J., Sinitskii, A., Abramova, V., Klimonsky, S., Tretyakov, Y. (2007). Directional emission from rare earth ions in inverse photonic crystals. *Applied Physics B*, Vol.89, No2-3 (November 2007), pp. 251-255, ISSN 0946-2171

Lin, Z., Vuckovic, Je. (2010). Enhanced two-photon processes in single quantum dots inside photonic crystal nanocavities. *Physical Review B*, Vol. 81, No 035301, (January 2010), pp. 1-5, ISSN 1098-0121

Moiseyenko, V., Dergachov, M., Burak, Ya. (2006). The influence of point defects on the vibrational spectrum of $Li_2B_4O_7$ crystal. *Ukrainian Journal of Physical Optics*, Vol.7, No.1, (March 2006), pp. 35-40, ISSN 1609-1833

Moiseyenko, V., Guziy, O., Gorelik, V., Dergachov, M. (2008). Optically excited secondary emission spectra of photonic crystals based on synthetic opals. *Optics and spectroscopy*, Vol.105, No.6, (December 2008), pp. 919-923, ISSN 0030-400X

Moiseyenko, V., Dergachov, M., Shvachich, V., Yevchik, A. (2009a). On the nature of the secondary emission of globular SiO_2 photonic crystals, *Proceedings of Metamaterials 2009 3rd International Congress on Advanced Electromagnetic Materials in Microwaves and Optics*, pp. 638-640, ISBN 978-0-9551179-6-1, London, UK, August 30 - September 4, 2009

Moiseyenko, V., Dergachov, M., Shvachich, V., Yevchik, A. (2009b). The possibility for surface-enhanced Raman scattering and spontaneous parametric down-conversion by globular photonic crystals infiltrated with dielectrics. *Ukrainian Journal of Physical Optics*, Vol.10, No.4, (December 2009), pp. 201-205, ISSN 1609-1833

Moiseyenko, V., Dergachov, M., Shvachich, V., Shvets, T., Roschenko, O. (2010). Spontaneous emission of laser dye molecules in synthetic opals under conditions of low dielectric contrast. *Ukrainian Journal of Physical Optics*, Vol.11, No.1, (March 2010), pp. 1-5, ISSN 1609-1833

Moiseyenko, V., Dergachov, M., Shvachich, V. (2012). Spontaneous parametric light scattering in spatially inhomogeneous nonlinear media based on photonic crystals *Optics and spectroscopy*, Vol.112, No.2 (February 2012), pp. 198-200, ISSN 0030-400X

Nasr, M., Carrasco, S., Saleh, B., Sergienko, A., Teich, M., Torres, J., Torner, L., Hum, D., Fejer, M. (2008). Ultrabroadband biphotons generated via chirped quasi-phase-matched optical parametric down-conversion. *Physical Review Letters*, Vol.100, No183601, (May 2008), pp. 1-4, ISSN 0031-9007

Podolskyy, D., Moiseyenko, V., Gorelik V., Shvachich V. (2006). Reflection spectra of bare and infiltrated with $Ba(NO_3)_2$ synthetic opal photonic crystals. *Ukrainian Journal of Physical Optics*, Vol.7, No.2, (April 2006), pp. 58-62, ISSN 1609-1833

Poulet, H., Mathieu, J.-P. (1970). *Spectres de Vibration et Symetrie des Cristaux*, Gordon & Breach Publishing Group, ISBN 0677501803, Paris - Londres - New York

Samarov, E., Mokrushin, F., Masalov, V., Abrosimova, G., Emel'chenko, G. (2006). Structural modification of synthetic opals during thermal treatment. *Physics of the Solid State*, Vol.48, No.7, (July 2006), pp. 1280-1283, ISSN 1063-7834

Stöber, W., Fink, A., Bohn, E. (1968). Controlled growth of monodisperse silica spheres in micron size range. *Journal of Colloidal and Interface Science*, Vol.26, No.2, (February 1968), pp. 62-68

Vats, N., John, S., Busch, K. (2002). Theory of fluorescence in photonic crystals. *Physical Review A*, Vol.65, No043808, (March 2002), pp. 1-13, ISSN 1050-2947

Vignolini, S., Riboli, F., Intonti, F., Belotti, M., Gurioli, M., Chen, Y., Colocci, M., Andreani, L., Wiersma, D. (2008). Local nanofluidic light sources in silicon photonic crystal microcavities. *Physical Review E*, Vol.78, No045603(R), (October 2008), pp. 1-4, ISSN 1539-3755

Yablonovitch, E. (1987). Inhibited spontaneous emission in solid-state physics and electronics. *Physical Review Letters*, Vol.58, No20, (May 1987), pp. 2059-2062, ISSN 0031-9007

6

Resonant Effects of Quantum Electrodynamics in the Pulsed Light Field

Sergei P. Roshchupkin, Alexandr A. Lebed', Elena A. Padusenko and Alexey I. Voroshilo
Institute of Applied Physics, NASU
Ukraine

1. Introduction

Studying of various aspects of laser field influence on physical processes is one of the most topical problems of modern applied and fundamental physics. Scientific interest is due to numerous unknown before phenomena, which are caused by laser radiation application and make enable coming to the main point of atomic and molecular structure of matter. These phenomena are of great importance over such fields of physics as holography, fiberglass optics, telecommunications, material authority, biophysics, plasma physics, nuclear fusion and so on. The lasers which generate radiation within the range from deep infrared to ultraviolet one and even the soft X-rays region with intensities up to 10^{22} W/cm^2 inclusive are made accessible at present. The sources of laser radiation had been put into practice of modern experiment widespread owing to its unique properties. The laser physics progress is generally concentrated on ever shorter and more powerful laser pulses production and on application of the lasts into various fields of scientific studies. New experimental conditions require continual improvements in computations and development of model of external field description.

Influence of laser field on kinematics and cross-sections of various quantum electrodynamics (QED) processes of the both first and second orders in the fine structure constant has been an object of study over a long period of time already. The characteristic feature of electrodynamics processes of the second order in the fine-structure constant in a laser field is associated with the possibility of their nonresonant and resonant modes. At this rate resonant cross-sections of scattering of particles may exceed the corresponding ones in external field absence in several orders of magnitude. Resonant character relates to the fact that lower-order processes, such as spontaneous emission or one-photon production and annihilation of electron-positron pairs, are allowed in the field of a light wave. Therefore, within a certain range of energy and momentum values a particle in an intermediate state may fall within the mass shell. Then the considered higher-order process effectively decomposes into two consecutive lower-order processes. Occurrence of resonances in a laser field is one of the fundamental problems of QED in strong fields.

Theoretical study of QED processes in an external laser field basis on solutions of the Dirac's equation for an electron in the field of a plane electromagnetic wave namely the Volkov functions (Volkov (1935)). Also one has to use the Green function of an intermediate particle in a plane wave field when studying processes of the second order in the fine structure constant. The analytical expression of the Green function was obtained (Schwinger (1951); Brown & Kibble (1964)).

Several of significant reviews are already devoted to studying of QED processes in the field of a plane monochromatic wave. The review Nikishov & Ritus (1979) is to be mentioned as one of the earliest works, in which first order processes in the field of a plane electromagnetic wave are studied generally. Processes of an electron scattered by an atom and a multiphoton ionization were considered in Ehlotzky et al. (1998). Theoretical studies of resonant and coherent effects of QED in light field were systematized in the monograph Roshchupkin & Voroshilo (2008) and several QED processes in strong field were reviewed by Ehlotzky et al. (2009).

Detailed consideration of resonant processes in the field of a plane monochromatic wave was fulfilled by Roshchupkin (1996). It is necessary to emphasize, that, when the resonance conditions are satisfied, the amplitude of process of particles scattering in the field of a plane monochromatic wave becomes infinite nominally. The infinity is eliminated by introducing of radiative corrections into Green's function of an intermediate particle according to the Breit–Wigner prescription under consideration as usual. The resonant peak altitude is determined by the lifetime of a particle in the intermediate state (Oleinik (1967)).

Since 1996 experiments of verification of QED in strong fields had been started at SLAC National Accelerator Laboratory (Bula et al. (1996); Burke et al. (1997)) along with theoretical study. The earliest results were related to studying of photon emission by an electron in a collision with laser pulse and photoproduction of electron–positron pairs by a gamma-quantum in the field of a laser. Verification of QED in strong pulsed fields is also expected in the frame of the wide-ranging FAIR project (Darmstadt, Germany). Within the FAIR project the laser facility PHELIX was developed and constructed. It enables to get laser pulses with power up to petawatt range. The earliest experiments at this laser facility have been put into practice (Bagnoud et al. (2009)).

The present paper reviews studies of a number of resonant processes in the field of an intense pulsed laser. The earliest studies on spontaneous bremsstrahlung of an electron in a collision with a laser pulse and photoproduction of electron–positron pairs by a high–energy photon in the pulsed field were performed by Narozhniy & Fofanov (1996). Second order processes in the pulsed fields which allow resonances were analytically studied for the case of moderately strong field (Lebed' & Roshchupkin (2010); Padusenko & Roshchupkin (2010); Lebed' & Roshchupkin (2011); Voroshilo et al. (2011)). These works were performed in recent several years therefore the systematization and generalization of them is definitely significant. It is important to underline that resonant divergences in amplitudes of studied processes are eliminated in a consistent manner due to account of a pulsed character of the external field in mentioned works.

Amplitude of a field strength of intense ultra short laser pulses changes greatly in space and time. Thus, description of the external field as a plane monochromatic wave becomes the problematic one. The external field is modeled usually as a plane quasi-monochromatic wave for description of interaction of particles with a pulsed laser field when the characteristic number of the external field oscillations in an electromagnetic pulse N is large:

$$N = \frac{\omega\tau}{2\pi} \gg 1, \tag{1}$$

where ω is the characteristic frequency of wave field oscillation, τ is the characteristic pulse duration. Quantity τ can approach a value of even tens of femtoseconds for fields within the optical range of frequency, thus the condition (1) is satisfied for the majority of modern intense pulsed lasers. Fields are named the quasi-monochromatic ones when the condition (1) is satisfied.

Hereinafter we consider the external electromagnetic pulse as a plane electromagnetic elliptically polarized wave propagating along z-axis with the four-potential (Narozhniy & Fofanov (1996))

$$A(\varphi) = g\left(\frac{\varphi}{\omega\tau}\right) \cdot \frac{cF_0}{\omega}\left(e_x \cos\varphi + \delta e_y \sin\varphi\right), \quad \varphi = (kx) = \omega t - \mathbf{kx}, \tag{2}$$

where φ is the wave phase; c is the velocity of light in vacuum, F_0 is the strength of a wave electric field in a pulse peak, $k = (\omega/c, \mathbf{k})$ is the wave four-vector; δ is the wave ellipticity parameter ($\delta = 0$ corresponds to the linear polarization case, $\delta = \pm 1$ corresponds to the circular polarization case); $e_x = (0, \mathbf{e}_x)$, $e_y = (0, \mathbf{e}_y)$ are the wave polarization four-vectors meeting the standard conditions: $e_x^2 = e_y^2 = -1$, $(e_x k) = (e_y k) = 0$. The function $g(\varphi/\omega\tau)$ is the envelope function of the external wave four-potential that allows to take into account the pulsed character of a laser field. Generally it is chosen to be equal to unity in the center of a pulse and to decrease exponentially when $|\varphi| \gg \omega\tau$. Thus, in this case it is possible to consider the parameter τ as the laser pulse characteristic duration.

Nonlinear effects in the processes of interaction of particles with the field of wave are governed by the classical relativistic-invariant parameter

$$\eta_0 = \frac{eF_0\lambda\!\!\!^{-}}{mc^2}. \tag{3}$$

Its value equals to the ratio of work done by the field at the wavelength to the particle rest energy. The parameter (3) is one of the most important characteristics of the external field and means the velocity of particle oscillation in the field of a wave in the case $\eta_0 \ll 1$. Multiphoton processes occurring when particles interact in a light field are characterized also by the Bunkin–Fedorov quantum parameter (Bunkin & Fedorov (1966))

$$\gamma_0 = \eta_0 \frac{mvc}{\hbar\omega}. \tag{4}$$

In the Eqs. (3), (4) e and m are the electron charge and mass, respectively; $\lambda\!\!\!^{-} = c/\omega$ is the characteristic wavelength, v is the particle velocity. The multiphoton parameters (3), (4) vary considerably with external field intensity. Thus, within the range of optical frequencies ($\omega \sim 10^{15}\ \mathrm{s}^{-1}$) the classical parameter $\eta_0 \sim 1$ when $F_0 \sim 10^{10} \div 10^{11}$ V/cm, and $\gamma_0 \sim 1$ when $F_0 \sim \left(10^5 \div 10^6\right)(c/v)$ V/cm. Consequently, we study all the processes within the range of moderately strong field when the considered parameters obey the following conditions:

$$\eta_0 \ll 1, \quad \gamma_0 \gtrsim 1. \tag{5}$$

The relativistic system of units, where $\hbar = c = 1$ and the standard metric for 4-vectors $(ab) = a_0 b^0 - \mathbf{ab}$ will be used throughout this paper.

2. Resonant spontaneous bremstrahlung of an electron scattered by a nucleus in the field of a pulsed light wave

We consider in this section the problem of spontaneous bremsstrahlung (SB) of an electron scattered by a nucleus in the external field of a pulsed light wave (see Fig. 1). Studying of SB when an electron is scattered by a nucleus or by an atom in presence of an external electromagnetic field has had a long-standing interest. Analytic expressions for the radiation spectrum of SB in a plane monochromatic wave in the nonrelativistic case have been derived by Karapetian & Fedorov (1978) for any atomic potential field in the Born approximation and by Zhou & Rosenberg (1993) for a short-range potential in the low-frequency approximation. Resonant SB of a nonrelativistic electron scattered by a nucleus in a plane-wave field was studied by Lebedev (1972). Borisov et al. (1980) considered resonant SB, which accompanies collisions of ultrarelativistic electrons for the case of large transferred momenta. In the general relativistic case the problem of electron-nucleus SB in the field of a plane monochromatic wave was studied by Roshchupkin (1985). It should be noted that the theory of SB in presence of an external field is also developed in Lötstedt et al. (2007); Schnez et al. (2007). They contain important numeric calculations for the case of a strong field. Nonresonant SB in a pulsed field was considered by Lebed' & Roshchupkin (2009). Resonant SB of an electron scattered by a nucleus in the field of a pulsed light wave was studied in the general relativistic case (Lebed' & Roshchupkin (2010)).

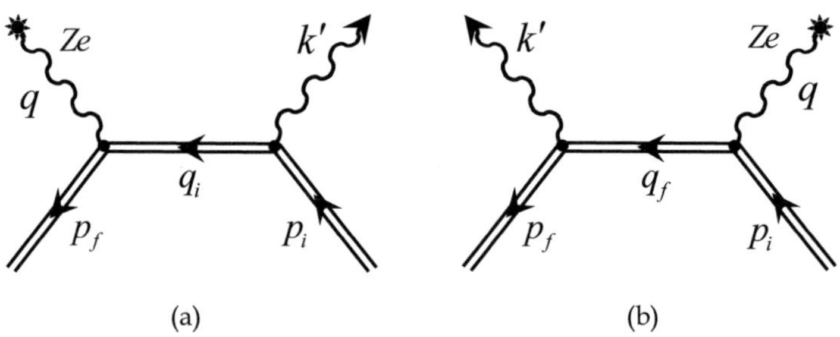

Fig. 1. Feynman diagrams of electron-nucleus SB in the field of a pulsed light wave. Incoming and outgoing double lines correspond to the Volkov functions of an electron in initial and final states; inner lines designate the Green function of an electron in a pulsed field. Wavy lines correspond to four-momenta of spontaneous photon and "pseudophoton" of nucleus recoil.

2.1 Amplitude of resonant spontaneous bremsstrahlung

The process of electron-nucleus SB in a pulsed light field (2) in the Born approximation on interaction of an electron with a nucleus, which corresponds to the transition of an electron from an initial state with the four-momentum $p_i = (E_i, \mathbf{p}_i)$ into a final state with the four-momentum $p_f = \left(E_f, \mathbf{p}_f\right)$, is described by two Feynman diagrams (Fig. 1).

The S-matrix element is given by

$$S_{fi} = -ie^2 \int d^4x_1 d^4x_2 \bar{\psi}_f (x_2 |A) \left[\tilde{\gamma}_0 A_0 (|\mathbf{x_2}|) G (x_2 x_1 |A) \hat{A}' (x_1, k') + \right. \\ \left. + \hat{A}' (x_2, k') G (x_2 x_1 |A) \tilde{\gamma}_0 A_0 (|\mathbf{x_1}|)\right] \psi_i (x_1 |A). \tag{6}$$

Here, $\psi_i (x_1|A)$ and $\bar{\psi}_f (x_2|A)$ are wave functions of an electron in initial and final states in the field (2), and $G (x_2 x_1 |A)$ is the Green function of an intermediate electron in the field of a pulsed light wave (2). Hereafter, expressions with hats above mean scalar products of corresponding four-vectors with the Dirac $\tilde{\gamma}$-matrices. In the amplitude (6) $A_0 \left(\left|\mathbf{x}_j\right|\right)$ is the Coulomb potential of a nucleus, and $A'_\mu \left(x_j, k'\right)$ is the four-potential of a spontaneously radiated photon. They have the following forms

$$A_0 \left(\left|\mathbf{x}_j\right|\right) = \frac{Ze}{\left|\mathbf{x}_j\right|}, \tag{7}$$

$$A'_\mu \left(x_j, k'\right) = \sqrt{\frac{2\pi}{\omega'}} \varepsilon^*_\mu \exp\left(ik'x_j\right), \quad j = 1, 2. \tag{8}$$

Here, ε^*_μ and $k' = (\omega', \mathbf{k}')$ are the polarization four-vector and the four-momentum of a spontaneous photon, $k'x_j = \omega' t_j - \mathbf{k}'\mathbf{x}_j$.

The SB amplitude of an electron scattered by a nucleus in the field of a moderately strong pulsed wave (6) in the general relativistic case was derived early (Lebed' & Roshchupkin (2009)). This amplitude may be presented in the following form

$$S_{fi} = \sum_{l=-\infty}^{\infty} S_l, \tag{9}$$

where S_l is the process partial amplitude with emission or absorption of $|l|$ laser-wave photons, that is

$$S_l = -i \frac{Ze^3\sqrt{\pi}}{\sqrt{2\omega' E_f E_i}} \bar{u}_f \left[B_{li} (\tilde{\gamma}_0, \hat{\varepsilon}^*) + B_{lf} (\hat{\varepsilon}^*, \tilde{\gamma}_0)\right] u_i. \tag{10}$$

Here, the functions $B_{li} (\tilde{\gamma}_0, \hat{\varepsilon}^*)$ and $B_{lf} (\hat{\varepsilon}^*, \tilde{\gamma}_0)$ correspond to the diagrams of electron-nucleus SB in Fig. 1; u_i, $\bar{u}_f$ are the Dirac bispinors.

Let us consider the diagram (a):

$$B_{li} (\tilde{\gamma}_0, \hat{\varepsilon}^*) = \sum_{r=-\infty}^{\infty} \frac{2\omega\tau^2}{\mathbf{q}^2 + q_0 (q_0 - 2q_z)} \int_{-\infty}^{\infty} d\xi \frac{\Lambda_{l+r} (\xi) \left[\hat{q}_i + m + \xi\hat{k}\right] \Lambda_{-r} (\xi)}{q_i^2 - m^2 + 2\xi (kq_i) + i0}, \tag{11}$$

where the four-vector $q = (q_0, \mathbf{q})$ is the transferred four-momentum, q_i is the four-momentum of an intermediate electron for the diagram (a) (Fig. 1)

$$\begin{cases} q = p_f - p_i + k' + lk, \\ q_i = p_i - k' + rk, \\ q_f = p_f + k' + (l + r) k; \end{cases} \tag{12}$$

q_f is the four-momentum of an intermediate electron for the diagram (b) (Fig. 1). The integral functions Λ_{l+r}, Λ_{-r} are specified as

$$\begin{cases} \Lambda_{l+r}(\xi) = \tilde{\gamma}_0 \int\limits_{-\infty}^{\infty} d\phi \cdot L_{l+r}(\phi) \cdot \exp\{iq_0\tau\phi - i(\xi\omega\tau)\phi\}, \\ \Lambda_{-r}(\xi) = \int\limits_{-\infty}^{\infty} d\phi' \cdot F_{-r}(\phi') \cdot \exp\{i(\xi\omega\tau)\phi'\}. \end{cases} \tag{13}$$

The integration variables in Eqs. (13):

$$\phi = \frac{\varphi}{\omega\tau}, \quad \phi' = \frac{\varphi'}{\omega\tau}. \tag{14}$$

The integral functions $F_{-r}(\phi')$, $L_{l+r}(\phi)$ in Eqs. (13) are stepless depended on the integration variables (14), and are determined as

$$\begin{aligned} F_{-r}(\phi') = \hat{\varepsilon}^* \cdot L_{-r}\left(\chi_{q_i p_i}, \gamma_{q_i p_i}(\phi'), \beta_{q_i p_i}(\phi')\right) + \\ + (e_+ b) \cdot g(\phi') L_{-r+1}\left(\chi_{q_i p_i}, \gamma_{q_i p_i}(\phi'), \beta_{q_i p_i}(\phi')\right), \end{aligned} \tag{15}$$

where

$$b = \frac{1}{4}\eta_0 m \left(\frac{\hat{\varepsilon}^* \hat{k} \tilde{\gamma}}{(kp_i)} + \frac{\tilde{\gamma} \hat{k} \hat{\varepsilon}^*}{(kq_i)} \right), \tag{16}$$

$$e_+ = e_x + i\delta e_y, \tag{17}$$

$$\begin{aligned} & L_{-r}\left(\chi_{q_i p_i}, \gamma_{q_i p_i}(\phi'), \beta_{q_i p_i}(\phi')\right) = \\ & = \frac{1}{2\pi} \int\limits_0^{2\pi} d\varphi \exp\left\{ i\left[\gamma_{q_i p_i}(\phi') \sin(\varphi - \chi_{q_i p_i}) + \beta_{q_i p_i}(\phi') \sin 2\varphi + r\varphi \right] \right\}. \end{aligned} \tag{18}$$

The arguments of functions (18) are defined by the expressions

$$\tan \chi_{q_i p_i} = \delta \frac{(e_y Q_{q_i p_i})}{(e_x Q_{q_i p_i})}, \quad Q_{q_i p_i} = \frac{q_i}{(kq_i)} - \frac{p_i}{(kp_i)}, \tag{19}$$

$$\gamma_{q_i p_i}(\phi') = \eta_0 g(\phi') \cdot m \sqrt{(e_x Q_{q_i p_i})^2 + \delta^2 (e_y Q_{q_i p_i})^2}, \tag{20}$$

$$\beta_{q_i p_i}(\phi') = \frac{1}{8}\left(1 - \delta^2\right) \eta_0^2 g^2(\phi') m^2 \left[\frac{1}{(kq_i)} - \frac{1}{(kp_i)} \right]. \tag{21}$$

Expressions for integral functions $L_{l+r}(\phi) \equiv L_{l+r}\left(\chi_{p_f q_i}, \gamma_{p_f q_i}(\phi), \beta_{p_f q_i}(\phi)\right)$ may be easily obtained from the appropriate expressions (18)-(21) after following replacements of indices and four-momenta: $-r \to l + r$, $q_i \to p_f$, $p_i \to q_i$.

Functions $L_n(\chi, \gamma, \beta)$ determine probabilities of multiphoton processes produced by the presence of a strong external field. Note that properties of these functions were studied by Roshchupkin et al. (2000) in detail. Thus, they may be represented as series in integer-order Bessel functions, i.e.

$$L_n(\chi, \gamma, \beta) = \exp(-in\chi) \sum_{s=-\infty}^{\infty} \exp(2is\chi) \cdot J_{n-2s}(\gamma) \cdot J_s(\beta). \tag{22}$$

The form of integral functions (18) is considerably simplified for the case of a circular polarization of an external light wave:

$$L_{-r}\left(\chi_{q_i p_i}, \gamma_{q_i p_i}\left(\phi'\right), 0\right) = \exp\left(ir\chi_{q_i p_i}\right) \cdot J_{-r}\left(\gamma_{q_i p_i}\left(\phi'\right)\right). \tag{23}$$

It is obvious from Eqs. (11), (13) that the essential range of the integration variable ξ is determined by the condition

$$|\xi| \lesssim \frac{1}{\omega\tau} \ll 1. \tag{24}$$

In view of quick oscillation of the integrand under $|\xi| \gg (\omega\tau)^{-1}$ the integrals in Eqs. (13) are small. Note that the expression of the amplitude $B_{lf}\left(\hat{\varepsilon}^*, \tilde{\gamma}_0\right)$ may be easily obtained from Eqs. (11), (13)-(21), if the replacements $q_i \to q_f$, $\tilde{\gamma}_0 \leftrightarrow \hat{\varepsilon}^*$ will be performed.

We emphasize, that dependence of the integrand denominator in Eq. (11) on the integration variable expresses consequence of accounting of the field pulsed character. The similar correction is absent in the monochromatic wave case, thus the resonant infinity in the amplitude of SB of an electron scattered by a nucleus in a light field occurs.

2.2 Resonance conditions

Fulfillment of the energy-momentum conservation law for components of a process of the second order caused a phenomenon when a virtual intermediate particle becomes real – that is, on-shell. Thus, the resonant divergence occurs in the process's amplitude. The energy-momentum conservation law for QED processes in a pulsed light field does not fulfill strictly. This peculiarity is inessential when nonresonant processes are studied. On the contrary, the energy-momentum nonconservation in the case of resonant SB of an electron scattered by a nucleus in a pulsed light field results following resonance conditions

$$q_j^2 - m^2 \lesssim \frac{\left(kq_j\right)}{\omega\tau}, \quad j = i, f. \tag{25}$$

(it follows from consideration of Eqs. (11), (24)). Therefore, the four-momentum of an intermediate electron occurs near the mass shell.

It is convenient to set down expressions which determine $q_{i,f}$ and q (12) in following form for the both amplitudes (a) and (b) (Fig. 1), respectively

$$\begin{cases} p_i + rk = q_i + k', \\ q = p_f - q_i + (l + r)\,k; \end{cases} \tag{26}$$

$$\begin{cases} q_f + rk = p_f + k', \\ q = q_f - p_f + (l + r)\,k. \end{cases} \tag{27}$$

Eqs. (26)-(27) represent the four-momentum conservation laws for the diagrams' vertices. These laws are fulfilled for only the values $r > 0$ under the conditions (25).

It is easy to ascertain that if a spontaneous photon propagates in the same direction as a photon of an external field, the conditions (25) cannot be satisfied simultaneously with the conservation laws (26) or (27) because the transit amplitude equals zero in this case. Therefore, resonances occur only when these photons propagate nonparallel to each other.

Taking Eq. (25) into account, we can use Eqs. (26), (27) for a moderately strong field (5) to find the frequency of a spontaneous photon in the resonance (the resonant frequency) for the both direct and exchange amplitudes (Figs. 1(a) and 1(b), respectively). Within zeroth order with respect to the small parameter $(\omega\tau)^{-1}$ the resonant frequency is specified:

$$\omega'_{res} \equiv \omega'_j = r\omega_j \frac{1}{1 \pm d_j}, \quad j = i, f, \tag{28}$$

where the signs "+" and "–" correspond to index values i and f, respectively,

$$\omega_j = \omega \cdot \frac{\kappa_j}{\kappa'_j}, \quad d_j = r\left(nn'\right) \cdot \frac{\omega}{\kappa'_j}, \tag{29}$$

$$\kappa_j = E_j - \mathbf{n}\mathbf{p}_j, \quad \kappa'_j = E_j - \mathbf{n}'\mathbf{p}_j, \tag{30}$$

$$n = \frac{k}{\omega} = \left(1, \mathbf{n}\right), \quad n' = \frac{k'}{\omega} = \left(1, \mathbf{n}'\right). \tag{31}$$

It is obvious from Eq. (29), that within a rather broad range of electron energies and scattering angles we have $d_j \ll 1$ (except an ultrarelativistic electron with the energy $\sim m^2/\omega$, moving within a narrow cone close to the direction of the momentum of a spontaneous photon). Therefore, resonances are mainly observed when the frequency of a spontaneous photon is multiple to ω_j (29).

Eqs. (28)-(31) for the resonant frequency imply that we may separate four characteristic domains of the frequency ω_j: the nonrelativistic case, $\omega_j \cong \omega$; the limiting case of ultrarelativistic energies, when an electron moves within a narrow cone related to a photon of an external field $\omega_j \ll \omega$; an ultrarelativistic electron moves within a narrow cone with a spontaneous photon, $\omega_j \gg \omega$; otherwise, $\omega_j \sim \omega$. Here, we consider resonant frequencies in detail.

The four-momentum conservation law (26) and the function F_{-r} explicit form (15) result that this function represents the amplitude of such process: an electron with the four-momentum p_i absorbs r photons of the external wave and emits a photon with four-momentum k'. This process was considered by Nikishov & Ritus (1979) in the case of a plane monochromatic wave, and by Narozhniy & Fofanov (1996) in the case of a pulsed light wave. The expression for the transferred four-momentum q (see the second equality in Eq. (26)) shows that the quantity $L_{l+r}\left(\chi_{p_f q_i}, \gamma_{p_f q_i}(\phi), \beta_{p_f q_i}(\phi)\right)$ defines the amplitude of scattering of an intermediate electron with the four-momentum q_i by a nucleus in the field of a light wave with absorption or emission of $|l+r|$ wave photons. In the nonrelativistic limiting case this process was studied by Bunkin & Fedorov (1966). Denisov & Fedorov (1967) considered this process in the general relativistic case. The process when an electron scattered by a nucleus in a pulsed light wave was studied by Lebed' & Roshchupkin (2008).

Consequently, if the interference between the direct and the exchange amplitudes is absent, the process of resonant electron-nucleus SB in the field of a light wave effectively decomposes into two consecutive processes of the first order: emission of a photon with the four-momentum k' by an electron in a pulsed light wave and scattering of an electron by a nucleus in a pulsed light wave (see Fig. 2).

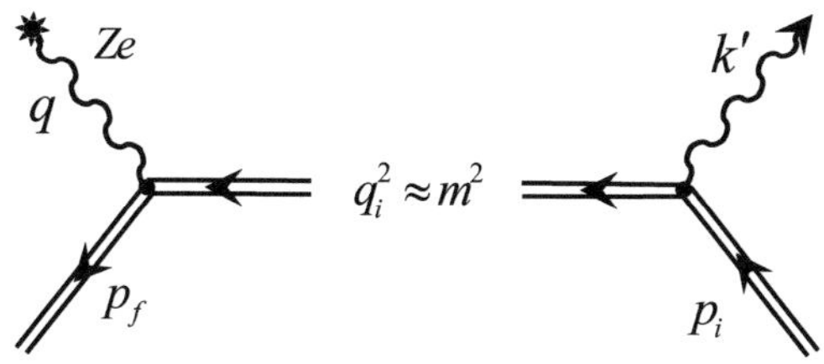

Fig. 2. Resonant electron-nucleus SB in the field of a pulsed light wave.

The difference for the other diagram (Fig. 1(b)) is concluded in the both replacement of the intermediate electron four-momentum ($q_i \to q_f$) and interchange of sequence of first order processes. Thus, an electron is scattered by a nucleus with absorption or emission of r wave photons, and then it spontaneously emits a photon with the four-momentum k' with $|l+r|$ wave photons absorption.

As it was pointed above, the integral functions (18) are determined by the integer-order Bessel functions (23) for the case of a circularly polarizated external wave. It is not difficult to verify that for given type polarization under the resonance conditions the arguments of the Bessel functions (20) may be represented as

$$\gamma_{q_i p_i}\left(\phi'\right) = 2r \cdot \eta_0 g\left(\phi'\right) \cdot \sqrt{\frac{u}{u_r} \cdot \left(1 - \frac{u}{u_r}\right)}. \tag{32}$$

Here, u, u_r are the relativistic invariant parameters

$$u = \frac{(kk')}{(kq_i)}, \quad u_r = 2r \cdot \frac{(kp_i)}{m^2}. \tag{33}$$

Equations (32)-(33) imply

$$\gamma_{q_i p_i}\left(\phi'\right) \sim \eta_0 \ll 1. \tag{34}$$

Consequently, within the range of fields specified by Eq. (5) the first resonance, that is, the resonance with $r = 1$, provides the main contribution to the resonant cross section, when the Bessel function has the largest value. This implies that the Compton scattering of a light wave by an initial electron is mainly due to absorption of one photon of an external field. However, the argument of the Bessel function $J_{l+r}\left(\gamma_{p_f q_i}(\phi)\right)$ is of the order of magnitude: $\gamma_{p_f q_i}(\phi) \sim \gamma_0 \gtrsim 1$. Thus, scattering of an intermediate electron by a nucleus in a pulsed wave field under these conditions is generally a multiphoton process.

Interference of the resonant amplitudes (which correspond to direct and exchange diagrams) implies the equality of their resonant frequencies, i.e. $\omega_i' = \omega_f'$. Within the field range specified by Eq. (5) the condition of interference between direct and exchange resonant amplitudes is written as:

$$\left(\mathbf{v}_f - \mathbf{v}_i\right)\left(\mathbf{n} - \mathbf{n}'\right) + \left(\mathbf{v}_f \times \mathbf{v}_i\right)\left(\mathbf{n}' \times \mathbf{n}\right) = \left(\mathbf{n}'\mathbf{n}\right) \cdot \frac{r\omega \cdot \left(\kappa_i + \kappa_f\right)}{E_i E_f}. \tag{35}$$

Here, $\mathbf{v}_j = \mathbf{p}_j/E_j$ is the electron velocity before $(j = i)$ and after $(j = f)$ scattering. The quantity involved in the right-hand side of Eq. (35) is small compared with the unity. Therefore, this equality is satisfied when directions of motion of photons (a spontaneous photon and a photon of an external field) or electrons (before and after scattering) are close to each other. It follows from Eq. (35) and from the fact that resonances vanish, when direction of spontaneous photon motion is close to direction of external field photon motion, that resonant amplitudes, which correspond to the processes shown on Figs. 1(a) and 1(b), interfere when an electron is scattered on the small angles, i.e.

$$\theta = \angle\left(\mathbf{v}_{i,}\mathbf{v}_f\right) \sim (1 - \mathbf{n}\mathbf{v}_i) \cdot (\omega/|\mathbf{v}_i|\, E_i) \ll 1. \tag{36}$$

Hereinafter, we consider the resonance of one diagram. We assume that the spontaneous photon frequency is equal

$$\omega' \approx \omega'_{res} = \omega'_i. \tag{37}$$

2.3 Amplitude integration

Let us study the process of resonant SB of an electron scattered by a nucleus in a pulsed light field at the expense of only one photon absorption, i.e. $r = 1$. The condition (24) allows to simplify the integration in Eq. (11)

$$\int_{-\infty}^{\infty} d\xi \frac{\exp\left\{i\xi\omega\tau\left(\phi' - \phi\right)\right\}}{q_i^2 - m^2 + 2\xi\left(kq_i\right) + i0} = \frac{\exp\left\{-2i\beta\left(\phi' - \phi\right)\right\}}{2\left(kq_i\right)} i\pi\left(sgn\left(\phi' - \phi\right) - 1\right). \tag{38}$$

Eq. (38) contains the relevant parameter, which determines resonant electron-nucleus SB in the field of a pulsed light wave:

$$\beta = \frac{q_i^2 - m^2}{4\left(kq_i\right)}\omega\tau. \tag{39}$$

As it can be seen from Eq. (39), values of the parameter β are defined by process kinematics and external pulsed-wave properties. This parameter specifies how closely the four-momentum of an intermediate electron coincides with the value on the mass shell under resonance conditions for electron-nucleus SB in the field of a pulsed light wave.

The subsequent analysis will be performed for the particular form of the envelope function of the pulsed light wave four-potential. We choose the Gaussian function:

$$g\left(\frac{\varphi}{\omega\tau}\right) = \exp\left\{-\left(\frac{2\varphi}{\omega\tau}\right)^2\right\} = \exp\left\{-\left(2\phi\right)^2\right\}. \tag{40}$$

Under the condition (34) the function $F_{-r}\left(\phi'\right)$ (15) in the amplitude may be expanded in the Taylor series. We may keep only linear terms with respect to the parameter η_0. For the envelope function (40), after simple computation we obtain the amplitude of resonant SB of an electron scattered by a nucleus in a pulsed light field:

$$B_{li}\left(\tilde{\gamma}_0, \hat{\varepsilon}^*\right) = \frac{2\pi \cdot \tilde{\gamma}_0\left(\hat{q}_i + m\right)\hat{F}}{\mathbf{q}^2 + q_0\left(q_0 - 2q_z\right)} \cdot \frac{-i\omega\tau\sqrt{\pi}}{4\left(kq_i\right)} \exp\{-\frac{\beta^2}{4}\} \cdot I\left(q_0, \beta\right), \tag{41}$$

$$\hat{F} = -\frac{1}{2}\exp\{i\chi_{q_ip_i}\} \cdot \gamma_{q_ip_i}(0) \cdot \hat{\varepsilon}^* + (e_+b). \quad (42)$$

$$I(q_0, \beta) = \tau \int_{-\infty}^{\infty} d\phi \cdot J_{l+1}(\phi) \exp\{i(q_0\tau + 2\beta)\phi\} \left(\operatorname{erf}\left(2\phi + \frac{i\beta}{2}\right) + 1\right). \quad (43)$$

Here, the function erf $(2\phi + i\beta/2)$ is the error function.

2.4 Cross-section of spontaneous bremsstrahlung

Let us calculate the differential probability during the entire time of electron-nucleus SB in a pulsed light field from the amplitude, Eqs. (9)-(10), (41)-(43) in standard manner (see Berestetskii et al. (1982)) for the spontaneous photon frequency (37):

$$dw = \sum_{l=-\infty}^{\infty} dw_l, \quad (44)$$

$$dw_l = \frac{Z^2 e^6 \pi}{2\omega' E_f E_i} \cdot \left|\bar{u}_f B_{li} u_i\right|^2 \cdot \frac{d^3 p_f d^3 k'}{T(2\pi)^6}. \quad (45)$$

Here, T is some comparatively large ($T \gtrsim \tau$) interval of the observation time. The energy-momentum conservation law for SB of an electron scattered by a nucleus in a pulsed light field does not fulfill strictly, however, under the condition (1) the essential range of integration is converged. Energies of a final electron are negligibly differ from the values, which are specified by the strict energy conservation law. We exclude small scattering angles from the consideration:

$$\theta = \angle\left(\mathbf{p}_i, \mathbf{p}_f\right) \gg \sqrt{\frac{\omega}{|\mathbf{p}_i|(\omega\tau)}} \ll 1. \quad (46)$$

The differential cross section of SB of an electron scattered by a nucleus in the field of a pulsed light wave is obtained by means of division of the probability per unit time per a flux density of scattered particles $v_i = |\mathbf{p}_i| / E_i$. Thus, we derive

$$d\sigma = \sum_{l=-\infty}^{\infty} d\sigma_l, \quad (47)$$

where $d\sigma_l$ is the partial cross section of a process with a spontaneous photon in the frequency interval $d\omega'$ within the solid angle $d\Omega'$, and a final electron within the solid angle $d\Omega_f$ with emission ($l > 0$) or absorption ($l < 0$) of wave photons. It may be written in the form

$$\frac{d\sigma_l}{d\omega' d\Omega' d\Omega_f} = \frac{Z^2 e^6 \omega' \pi (\omega\tau)^2}{(2\pi)^2 \mathbf{q}^4} \frac{\left|\mathbf{p}_f\right|}{|\mathbf{p}_i|} \frac{\exp\{-\beta^2/2\}}{64(kq_i)^2} \times$$
$$\times \left|\bar{u}_f M_i u_i\right|^2 \cdot \frac{\tau}{T} \int_{-T/2\tau}^{T/2\tau} d\phi \cdot J_{l+1}^2(\phi) \left|\operatorname{erf}\left(2\phi + \frac{i\beta}{2}\right) + 1\right|^2, \quad (48)$$

$$M_i = \tilde{\gamma}_0(\hat{q}_i + m)\hat{F}. \quad (49)$$

It is taken into account that $d^3p_f = \left|\mathbf{p}_f\right| E_f dE_f d\Omega_f$ and $d^3k' = \omega'^2 d\omega' d\Omega'$. It is important to note that the main contribution into the sum (47) is given by the processes with emission (absorption) of $|l| \lesssim \gamma_0$ number of wave photons within the range of a moderately strong field for electron relativistic energies $\left(E_{i,f} \gtrsim m\right)$. Therefore, the energy contribution of stimulated photons may be neglected ($|l|\,\omega/E_{i,f} \lesssim \eta_0 m/E_{i,f} \ll 1$) in Eq. (12). Thus, it is easy to sum over all possible partial processes of electron scattering by a nucleus (47).

If polarization effects are not of interest, then averaging over polarizations of an initial electron and summation over polarizations of a final electron and a spontaneous photon are to be done. Performing the relevant procedures of averaging and summation, we derive the general relativistic expression for the resonant differential cross section of electron-nucleus SB in a pulsed light field in the case of electron large-angle scattering (46)

$$\frac{d\sigma_{res}}{d\Omega'} = \frac{1}{\pi^2} \cdot \frac{E_i \kappa_i^2 \left|\mathbf{q}_i\right| u}{(nn')^2 \left|\mathbf{p}_i\right| (1+u)} \cdot P_{res} \cdot d\sigma_s dW^{(1)}. \tag{50}$$

Here,

$$d\sigma_s = 2Z^2 r_e^2 \frac{\left|\mathbf{p}_f\right| m^2}{\left|\mathbf{q}_i\right| \mathbf{q}^4} \left(m^2 + E_f q_{i0} + \mathbf{p}_f \mathbf{q}_i\right) d\Omega_f \tag{51}$$

is the differential cross section of scattering of an intermediate electron with the four-momentum q_i by a nucleus in a wave field; r_e is the classical electron radius.

$$dW^{(1)} = \frac{\alpha \eta_0^2 m^2}{4E_i} \left\{2 + \frac{u^2}{1+u} - \frac{4u}{u_1}\left(1 - \frac{u}{u_1}\right)\right\} \cdot \frac{du}{(1+u)^2} \tag{52}$$

is the probability that an electron with the four-momentum p_i absorbs one photon from an external field and emits a photon with the four-momentum k'. The function P_{res} in Eq. (50) has the form

$$P_{res} = \frac{\pi (\omega\tau)^2}{64 (kq_i)^2} \cdot P_{res}^{\beta}, \tag{53}$$

$$P_{res}^{\beta} = \exp\{-\beta^2/2\} \cdot \frac{1}{2\rho} \int\limits_{-\rho}^{\rho} d\phi \cdot \left|\operatorname{erf}\left(\phi + \frac{i\beta}{2}\right) + 1\right|^2, \tag{54}$$

$$\rho = T/\tau. \tag{55}$$

Here, the parameter ρ is the relation between the observation time and the pulse duration, its value is determined by conditions of the concrete experiment. Thus, if an external field is represented as electromagnetic pulses abiding one by one, then the parameter ρ assumes sense of the ratio of a distance between the nearest-neighbor pulses to the characteristic pulse duration. Dependence of the function P_{res} (53) on the parameter β (39) defines magnitude and shape of the resonant peak in the cross section of an electron-nucleus SB process in a pulsed light field (see Fig. 7).

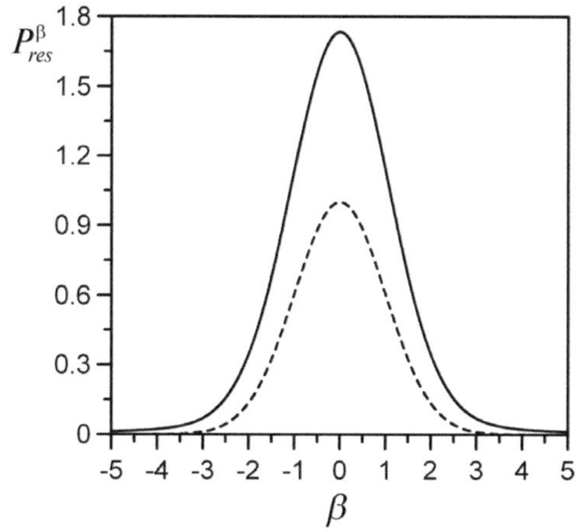

Fig. 3. Shape of the first resonant peak in the cross section of electron-nucleus SB in a pulsed light field ($\rho = 3$). The dashed line represents the Gaussian function: $\exp(-\beta^2/2)$.

2.4.1 Resonant width

In the frame of subsequent analysis we are to estimate the magnitude of the resonant width. For this purpose we consider the case when the four-momentum of an intermediate photon occurs near the mass shell:

$$\beta = \frac{(q_i^2 - m^2)}{4\,(kq_i)}\omega\tau \ll 1. \tag{56}$$

Thus, we can easily write

$$P_{res} \approx \frac{\pi}{4} \cdot \frac{(a_1/a_2)}{\left(q_i^2 - m^2\right)^2 + 4m^2\Gamma_\tau^2}, \tag{57}$$

where the width Γ_τ, caused by the pulsed character of an external wave, is equal to:

$$\Gamma_\tau = \frac{2}{\sqrt{a_2}} \frac{(kq_i)}{m} \frac{1}{\omega\tau}, \tag{58}$$

and the coefficients a_1 and a_2 are specified by

$$a_1 = \frac{1}{2\rho} \int_{-\rho}^{\rho} (\mathrm{erf}\,(\phi) + 1)^2\, d\phi, \tag{59}$$

$$a_2 = \frac{1}{2} - \frac{1}{4\sqrt{\pi}a_1\rho} \left(\sqrt{2}\mathrm{erf}\left(\sqrt{2}\rho\right) + \int_{-\rho}^{\rho} \phi \exp\left(-\phi^2\right) (\mathrm{erf}\,(\phi) + 1)\, d\phi \right). \tag{60}$$

It is important to underline that the relationship for the function (53) under the condition (56) turns into the standard resonant expression (57), which is usually used in the Breit-Wigner prescription. It is convenient to represent the resonant peak profile P_{res} in the form (57) to compare obtained results with corresponding ones for the case of a monochromatic wave. Note, that in the monochromatic wave case the resonant infinity in the cross section is eliminated by radiative corrections introducing into the Green function. The Breit-Wigner broadening prescription is concluded in addition of the imaginary part of the electron mass, that is $m \to m - i\Gamma_R$. Here, the radiation width is specified

$$\Gamma_R = \frac{1}{3}\alpha\eta_0^2 \frac{\sigma_c(q_i)}{\sigma_T} \cdot \frac{(kq_i)}{m}, \tag{61}$$

where $\sigma_c(q_i)$ is the total cross section of the Compton scattering of an external field photon by an intermediate electron with the four-momentum q_i (it is the most probable way out of an electron from an intermediate state), σ_T is the cross section of the Thomson scattering.

The resonant width (58) providing by finite time of particle-field interaction is so-called transit width. In real experiments the transit width value is generally determined by geometry of an experiment and linear sizes of space where a particle interacts with an external field. It can be seen from Eq. (58) that the transit width is specified by the pulse duration and process kinematics. Influence of the pulse duration on the resonant behavior of the electron-nucleus SB cross section was discussed by Schnez et al. (2007). The electromagnetic pulse duration has to be longer than the lifetime of an intermediate electron state. Otherwise, an electron will not have enough time to interact with a wave. Thus,

$$\tau \gtrsim \frac{1}{\Gamma_R}. \tag{62}$$

Requirements (62), (58) implies that values of the quantity $\omega\tau$ have to satisfy the following condition:

$$\omega\tau \gtrsim \frac{1}{\alpha\eta_0^2}\frac{\omega m}{(kq_i)}. \tag{63}$$

Comparison of the resonance widths for the pulse duration values (63) implies that $\Gamma_\tau \sim \Gamma_R$ within a sufficiently broad range of electron energies and scattering angles. Consequently, the both radiation and transit widths have to be simultaneously considered in resonant SB study. An exception is the case of ultrarelativistic energies when

$$\frac{1}{\alpha\eta_0^2}\frac{\omega m}{(kq_i)} \lesssim \omega\tau \ll \frac{1}{\alpha\eta_0^2}. \tag{64}$$

In this case $\Gamma_\tau \gg \Gamma_R$ and the expressions for the resonant differential cross section of electron-nucleus SB in a pulsed field (50)-(52), (57)-(60) are correct without radiation width accounting.

It should be pointed that we excluded other causes of the resonant peak widening from consideration. Thus, we assume that the Doppler broadening of the resonance (the real electron bunch is not monochromatic) and broadening caused by collisions of electrons are negligible.

2.4.2 Range of relativistic energies

In this section we consider the range of electron relativistic energies: $E_i \gtrsim m$. Here we eliminate the case when ultrarelativistic electrons are moving within a narrow cone with a spontaneous photon or an external field photon from consideration. Then $|d_i| \ll 1$ (it follows from Eq. (29)). Therefore, the resonant frequency ω_i' (28) in this case is of the order of the external field frequency. Depending on the spontaneous photon emission angle with respect to direction of the initial electron momentum the resonant frequency falls within the interval:

$$\omega \cdot \frac{\kappa_i}{E_i + |\mathbf{p}_i|} \leq \omega_i' \leq \omega \cdot \frac{\kappa_i}{E_i - |\mathbf{p}_i|}. \tag{65}$$

This frequency reaches its minimum and maximum when a spontaneous photon is emitted along direction of the initial electron motion and in opposite direction, respectively.

The invariant parameters (33) assume the form

$$u_r = 2r \cdot \frac{\omega \kappa_i}{m^2}, \quad u \cong (nn') \cdot \frac{\omega'}{\kappa_i} \ll 1. \tag{66}$$

Taking the radiation width into account, we may represent the resonant denominator (57) as

$$\left(q_i^2 - (m - i\Gamma_R)^2\right)^2 + (2m\Gamma_\tau)^2 = (2\omega' |\mathbf{p}_i|)^2 \cdot \left[\left(\cos\theta_i' - \cos\theta_{res}'\right)^2 + C_\tau^2\right]. \tag{67}$$

Here we introduced the notations

$$\theta_{i,f}' = \angle\left(\mathbf{k}', \mathbf{p}_{i,f}\right), \quad \theta_{i,f} = \angle\left(\mathbf{k}, \mathbf{p}_{i,f}\right), \tag{68}$$

$$\cos\theta_{res}' = \frac{E_i - (\omega/\omega') \cdot \kappa_i}{|\mathbf{p}_i|}, \quad C_\tau = \frac{m\Gamma_\tau \sqrt{1 + \mu_\tau^2}}{\omega' |\mathbf{p}_i|}, \tag{69}$$

$$\mu_\tau = \frac{\Gamma_R}{\Gamma_\tau} = \frac{\sqrt{a_2}}{6} \alpha \eta_0^2 \omega \tau. \tag{70}$$

For the resonant angles that are not too close to zero and π we can expand $\cos\theta_i'$ in Eq. (67) into the Taylor series near the resonant angle θ_{res}' with an accuracy up to the term of the first order with respect to $t = \theta_i' - \theta_{res}'$. The solid angle which corresponds to spontaneous photon emission is written as $d\Omega' = \sin\theta_{res}' d\varphi dt$. Then the resonant cross section (50) assumes the following form

$$d\sigma_{res} = \frac{1}{4\pi^2} \cdot \frac{d\varphi \cdot d(t/y)}{1 + (t/y)^2} \cdot \frac{E_i \kappa_i}{(nn') |\mathbf{p}_i| \Gamma_\tau \sqrt{1 + \mu_\tau^2 m}} \cdot dW^{(1)} d\sigma_S(q_i). \tag{71}$$

Here, $y = m\Gamma_\tau (1 + \mu_\tau)/(\omega' |\mathbf{p}_i| \sin\theta_{res}') \sim (\omega\tau)^{-1} \ll 1$. Since the resonance angular width is very small, we may integrate the expression (71) with respect to the azimuthal angle $d\varphi$, and with respect to $d(t/y)$ within the limits from zero to $+\infty$ (we extend the integration limits to infinity because of integral fast convergence). Finally, we derive

$$d\sigma_{res} = \frac{E_i \kappa_i}{2 (nn') m |\mathbf{p}_i| \Gamma_\tau \sqrt{1 + \mu_\tau^2}} \cdot dW^{(1)} d\sigma_S(q_i), \tag{72}$$

where

$$dW^{(1)} = \alpha \eta_0^2 (nn') \frac{m^2}{2E_i \kappa_i} \left\{1 - \frac{2u}{u_1} \cdot \left(1 - \frac{u}{u_1}\right)\right\} d\omega'. \tag{73}$$

Derived expressions (72)-(73) for the resonant cross section are valid within the range of field intensities (5) when an electron scatters into the large angles $\theta \gg \omega/|\mathbf{p}_i|$. Spontaneous photon frequency and emission angle with respect to the initial electron momentum are unambiguously related to each other by Eq. (69), where the spontaneous photon frequency is chosen from the interval (65).

Note, that the conventional cross section $d\sigma_*$ of electron-nucleus bremsstrahlung (in external field absence) may be factorized as a product of the cross section $d\sigma_S(p_i)$ of electron-nucleus elastic scattering (see (51)) and the probability dW_γ of photon emission

$$d\sigma_* = d\sigma_S \cdot dW_\gamma, \tag{74}$$

$$dW_\gamma = \frac{\alpha}{4\pi^2} \cdot \left\{ \mathbf{q}^2 - (\mathbf{n}'\mathbf{q})^2 \cdot \frac{m^2}{\kappa_i'\kappa_f'} \right\} \cdot \frac{d\omega'}{\omega'\kappa_i'\kappa_f'} \cdot d\Omega', \quad \mathbf{q} = \mathbf{p}_f - \mathbf{p}_i. \tag{75}$$

Let us calculate the ratio of the resonant cross section (72) to the conventional cross section of electron-nucleus bremsstrahlung (74) (in absence of an external field). At that we take into account the resonant relation (69) between spontaneous photon frequency and emission angle

$$R_{res} = \frac{d\sigma_{res}}{d\sigma_*/d\Omega'} = f_1 \cdot \pi^2 \eta_0^2 \frac{\omega\tau}{\sqrt{1+\mu_\tau^2}} \left(\frac{m}{|\mathbf{p}_i|} \right)^2, \tag{76}$$

where the function $f_1 \sim 1$ and has a rather cumbersome form:

$$f_1 = \frac{\sqrt{a_2}\kappa_f'}{2|\mathbf{p}_i|} \frac{1 - (nn')\frac{m^2}{\kappa_i\kappa_i'}\left(1 - (nn')\frac{m^2}{2\kappa_i\kappa_i'}\right)}{4\sin^2\left(\frac{\theta}{2}\right) - \left(\cos\theta_f' - \cos\theta_i'\right)^2 \frac{m^2}{\kappa_i'\kappa_f'}}. \tag{77}$$

Let us choose for calculation the laser field characteristic according to SLAC experiments (Bula et al. (1996)): laser-wave frequency, $\omega = 2.35$ eV; laser pulsewidth, $\tau = 1.5$ ps; field strength in a pulse peak, $F_0 = 6 \cdot 10^9$ V/cm; ratio between observation time and laser pulse width, $\rho = 5$. Fig. 4 displays the ratio of the resonant differential cross-section of electron-nucleus SB to the cross section of bremsstrahlung in absence of an external field (76) as a function of the electron velocity.

Eq. (76) and Fig. 4 show that within the range of electron relativistic energies the resonant differential cross section of electron-nucleus SB, when the scattered electron ejection angle is detected simultaneously with the spontaneous photon emission angle, may be five orders of magnitude greater than the corresponding cross section in external field absence. Within the range of electron ultrarelativistic energies this ratio decreases drastically: $R_{res} \sim (m/E_i)^2 \to 0$.

The ratio (76) as a function of the spontaneous photon azimuthal angle is of interest from a perspective of experimental testing of obtained results. In the actual experiment usually the radiation detection over the azimuthal angle is technically implemented significantly easier than over the polar angle. Fig. 5 displays $\lg R_{res}$ (76) as a function of the spontaneous photon azimuthal angle.

Fig. 5 shows that the ratio (76) may change its order of magnitude with the azimuthal angle value. This dependence is characterized by presence of two maxima in distribution over the azimuthal angle. Thus, when the final electron azimuthal angle coincides with the initial electron angle (it is scattering in the plane of the vectors $(\mathbf{k}, \mathbf{p}_i)$) the maxima in distribution

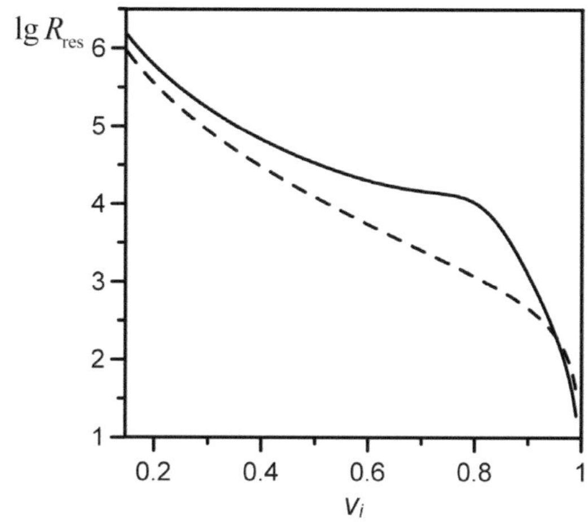

Fig. 4. Ratio R_{res} (76) as a function of the initial velocity for electron momentum preset orientations in initial ($\theta_i = 163°$, $\varphi_i = 0°$) and final ($\theta_f = 150°$,$\varphi_f = 0°$) states and spontaneous photon fixed orientation: solid line, $\theta' = 120°$, $\varphi' = 10°$; dashed line, $\theta' = 120°$, $\varphi' = 60°$.

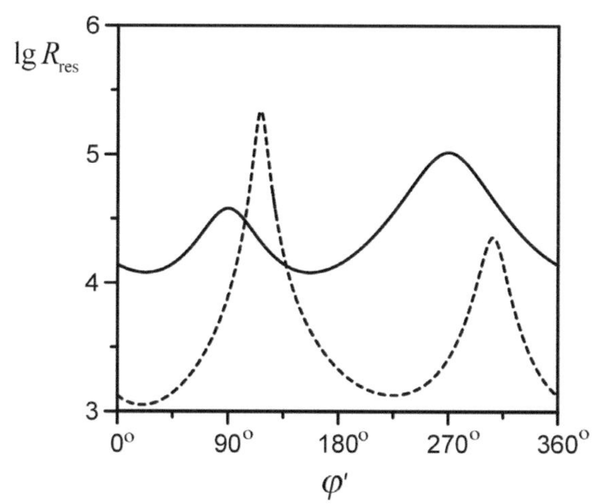

Fig. 5. Ratio R_{res} (76) as a function of the azimuthal angle of a spontaneous photon for electron momentum preset orientations in initial and final states and the spontaneous photon fixed polar angle: $\theta_i = 163°$, $\theta_f = 150°$, $\theta' = 120°$. Solid line, $\varphi_i = \varphi_f = 90°$; dashed line, $\varphi_i = 90°$, $\varphi_f = 320°$.

correspond to spontaneous photons emission just within this plane (solid line). In the case when a final electron scatters in another way the peak position in distribution over the azimuthal angle is specified by both initial and final azimuthal angles. The value of the ratio of the resonant differential cross section of electron-nucleus SB to the ordinary bremsstrahlung cross section as a function of the azimuthal angle may be changed in two orders of magnitude.

2.4.3 Range of nonrelativistic electron energies

In this section we assume that initial and final electron energies are small in comparison with the light speed: $Z\alpha \ll v_{i,f} \ll 1$. It follows from Eqs. (28)-(31) that resonant frequencies for nonrelativistic electrons are given by

$$\omega'_{i,f} = r\omega\left(1 + \mathbf{v}_{i,f}\left(\mathbf{n}' - \mathbf{n}\right)\right) \cong r\omega. \tag{78}$$

Thus, resonances occur when the spontaneous photon frequency is multiple to the external field frequency. The condition of interference between direct and exchange resonant amplitudes (35) is written as

$$\left(\mathbf{v}_f - \mathbf{v}_i\right)\left(\mathbf{n} - \mathbf{n}'\right) = 2r \cdot \left(nn'\right) \cdot \frac{\omega}{m} \ll 1, \tag{79}$$

and, consequently, interference appears when an electron scatters into the small angels $\theta \sim \omega/mv_i \ll 1$.

The resonant cross section in the case when a nonrelativistic electron scatters into the large angles is obtained from Eq. (50):

$$d\sigma_{res} = \frac{1}{2\left(nn'\right) v_i \Gamma_\tau \sqrt{1+\mu_\tau^2}} dW^{(1)} d\sigma_s, \tag{80}$$

where

$$dW^{(1)} = \frac{1}{2}\alpha\eta^2\left(nn'\right) \cdot \left\{1 - \frac{2u}{u_1} \cdot \left(1 - \frac{u}{u_1}\right)\right\} d\omega', \tag{81}$$

$$\frac{u}{u_1} = \left(nn'\right)\frac{\omega'}{2\omega},$$

$$d\sigma_s = (2Z)^2 r_e^2 \frac{\left|\mathbf{p}_f\right|}{\left|\mathbf{q}_i\right|}\frac{m^4}{\mathbf{q}^4} d\Omega_f. \tag{82}$$

The resonant frequency of a spontaneous photon depends on the emission angle of this photon with respect to the initial electron momentum and lies within a narrow interval:

$$\omega\left(1 - 2v_i \cos^2\left(\theta_i/2\right)\right) \le \omega'_{res} \le \omega\left(1 + 2v_i \sin^2\left(\theta_i/2\right)\right). \tag{83}$$

The transit width Γ_τ (58) and the radiation width Γ_R (61) in the nonrelativistic limit are given by

$$\Gamma_\tau = \frac{2}{\sqrt{a_2}}\frac{1}{\tau}, \quad \Gamma_R = \frac{1}{3}\alpha\eta^2\omega. \tag{84}$$

We may write the ratio of the resonant cross section (50) to the corresponding conventional nonrelativistic cross section of electron-nucleus bremsstrahlung as

$$R_{res} = f_2 \cdot \pi^2\eta_0^2 \frac{\omega\tau}{\sqrt{1+\mu_\tau^2}} v_i^{-3}, \tag{85}$$

where the function $f_2 \sim 1$ and has the form

$$f_2 = \frac{\sqrt{a_2}}{2} \frac{1 - (1/2)\sin^2\theta'}{4\sin^2(\theta/2) - \left(\cos\theta'_f - \cos\theta'_i\right)^2}. \tag{86}$$

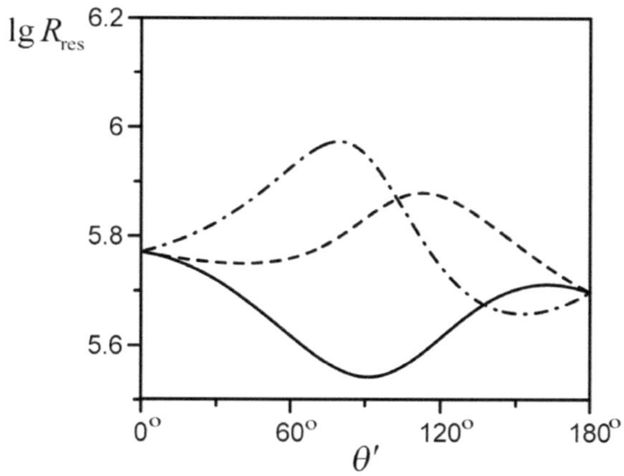

Fig. 6. Dependence of R_{res} (85) on the polar angle of a spontaneous photon for a nonrelativistic electron with the initial velocity $v_i = 0.1$. Orientations of the electron momentum in initial ($\theta_i = 163°$, $\varphi_i = 90°$) and final ($\theta_f = 150°$, $\varphi_f = 320°$) states are fixed. The azimuthal angle, corresponding to emission of a spontaneous photon, is (solid line) $\varphi' = 60°$, (dashed line) $\varphi' = 160°$, and (dash-dotted line) $\varphi' = 270°$.

Fig. 6 shows the dependence of quantity R_{res} (85) on the polar angle of spontaneous photon emission for a nonrelativistic electron with the initial velocity $v_i = 0.1$. Fig. 6 shows that for the case of electron kinetic energies of several kiloelectronvolts the resonant differential SB cross section may be 5–6 orders of magnitude greater than the corresponding cross section of bremsstrahlung in external field absence when the angle of spontaneous photon emission is detected simultaneously with the ejection angle of an electron scattered into the large angle.

2.4.4 Range of ultrarelativistic energies of electrons moving within a narrow cone with a photon from the wave

In this section we consider an ultrarelativistic electron that moves (in initial or final states) within the narrow cone related to an external field photon. Therefore, the quantities $\kappa_{i,f}$ (30) in Eqs. (28)-(31) may be written as

$$\kappa_{i,f} = \left(1 + \delta^2_{i,f}\right) \cdot m^2/2E_{i,f}, \quad \delta_{i,f} = \theta_{i,f} \cdot E_{i,f}/m. \tag{87}$$

Taking these relations into account and using Eqs. (28)-(31) we find that the resonant frequencies are much less than the external field frequency. They are given by:

$$\omega'_{i,f} = r\omega_{i,f}, \quad \omega_{i,f} = \frac{\left(1 + \delta^2_{i,f}\right)}{2(nn')} \cdot \left(\frac{m}{E_{i,f}}\right)^2 \cdot \omega \ll \omega. \tag{88}$$

From Eq. (88) follows that the condition of interference between direct and exchange resonant amplitudes implies that $\delta_i = \delta_f$ and $\theta_i \cong \theta_f$, that is, initial and final electrons form the equal angles with the external field photon momentum and are located on different sides of this photon momentum. Also, it can be seen from (35) that $\theta_i \sim \omega m^2/E_i^3 \ll 1$. When an ultrarelativistic initial electron moves within the narrow cone with an external field photon and scatters into the large angle $\theta_i \gg \omega m^2/E_i^3$ the resonant cross section is derived from Eq. (50) under the condition (87):

$$d\sigma_{res} = \frac{\left(1+\delta_i^2\right) m}{4\left(nn'\right) E_i \Gamma_\tau \sqrt{1+\mu_\tau^2}} \cdot dW^{(1)} \cdot d\sigma_S\left(q_i\right). \tag{89}$$

Here, the spontaneous photon resonant frequency is given by Eq. (88) with value $r = 1$, and the angle of spontaneous photon emission is not close to direction of initial electron motion. Ratio of the resonant cross section (89) to the conventional cross section of electron-nucleus bremsstrahlung may be derived from Eq. (76) with respect to Eq. (87):

$$R_{res} = f_3 \cdot \pi^2 \eta_0^2 \frac{\omega\tau}{\sqrt{1+\mu_\tau^2}} \left(\frac{m}{E_i}\right)^2, \tag{90}$$

where the function $f_3 \sim 1$ and has a rather cumbersome form.

It may be easily estimated that for the pulsed field parameters $\omega = 2.35$ eV, $\tau = 1.5$ ps, $F_0 = 6 \cdot 10^9$ V/cm, $\rho = 5$ and the electron energy $E_i = 5$ MeV the resonant cross section is of the order of the ordinary cross section when the angle of spontaneous photon emission is detected simultaneously with the ejection angle of an electron scattered on the large angle.

2.4.5 Range of ultrarelativistic energies of electrons moving within a narrow cone with a spontaneous photon

We suppose that an ultrarelativistic electron (an initial or a final one) moves within the narrow cone with a spontaneous photon. Then the quantities $\kappa'_{i,f}$ (30) may be written in an analogous to Eq. (87) form, where

$$\delta_{i,f} \to \delta'_{i,f} = \theta'_{i,f} \cdot E_{i,f}/m. \tag{91}$$

Here, depending on the electron energy we may deal with one of two possible situations. It is provided that $m \ll E_{i,f} \ll m^2/\omega$, than resonant frequencies fall within the interval $\omega \ll \omega'_{i,f} \ll E_{i,f}$ and are given by

$$\omega'_{i,f} = r\omega_{i,f}, \quad \omega_{i,f} = \frac{2\left(nn'\right)}{\left(1+\delta'^2_{i,f}\right)} \cdot \left(\frac{E_{i,f}}{m}\right)^2 \cdot \omega. \tag{92}$$

It was demonstrated by Roshchupkin (1985) that resonances do not occur for energies $E_{i,f} \gg m^2/\omega$. It is obviously that direct and exchange resonant amplitudes may interfere with each other only when initial and final electrons move within the narrow cone with a spontaneous photon, so $\delta'_i = \delta'_f$.

When an ultrarelativistic initial electron moves within the narrow cone with a spontaneous photon and scatters on the large angle $\theta \gg \omega/E_i$ we may use Eqs. (50)-(52) to find the resonant cross section. In this case, it is convenient to represent the resonant denominator in the following form

$$\left(q_i^2 - (m - i\Gamma_R)^2\right)^2 + (2m\Gamma_\tau)^2 = m^4\left[\left(x - \delta_i'^2\right)^2 + y^2\right] \cdot \frac{u^2}{(1+u)^2}, \tag{93}$$

where

$$x = \frac{u_1}{u} + \frac{(1+u)\,\Gamma_\tau^2\,(1+\mu_\tau^2)}{u \cdot m^2} - 1, \quad y = \frac{2\,(1+u)\,\Gamma_\tau\sqrt{1+\mu_\tau^2}}{u \cdot m}. \tag{94}$$

Here, the invariant parameters u, u_1 are given by

$$u \cong \frac{\omega'}{E_i - \omega'}, \quad u_1 = 2\left(nn'\right) \cdot \frac{\omega E_i}{m^2}. \tag{95}$$

Now Eqs. (50)-(52), (93) are to be taken into account, the solid angle is to be written as $d\Omega' = \left(m^2/2E_i^2\right) d\varphi d\delta_i'^2$, and integration should be performed with respect to the azimuthal angle, and $\delta_i'^2$ within the limits from zero to $+\infty$. Thus, we derive the following expression for the resonant cross section:

$$d\sigma_{res} = \Upsilon\,(xy) \cdot \frac{q_{i0}}{m\Gamma_\tau\sqrt{1+\mu_\tau^2}} \cdot dW^{(1)} \cdot d\sigma_S\,(q_i)\,. \tag{96}$$

Here,

$$\Upsilon\,(xy) = \frac{1}{\pi}\int_0^\infty \frac{d\delta_i'^2}{\left(x - \delta_i'^2\right)^2 + y^2} = \frac{1}{2} + \frac{1}{\pi} arctg\left(\frac{x}{y}\right) \tag{97}$$

is a smoothed step function. In regions far from the resonance $|u_1 - u| \gg 2\,(1+u)\,(\Gamma_\tau/m)$ and at the resonance point $u_1 = u$ this function takes the following limiting values:

$$\Upsilon\,(xy) = \begin{cases} 1, & \text{if } u < u_1, \\ 0.5, & \text{if } u = u_1, \\ y \cdot u/\pi\,(u - u_1)\,, & \text{if } u > u_1. \end{cases} \tag{98}$$

The probability is given by

$$dW^{(1)} = \alpha\eta^2 \cdot \frac{m^2}{4E_i} \cdot \left\{2 + \frac{u^2}{1+u} - \frac{4u}{u_1}\left(1 - \frac{u}{u_1}\right)\right\} \cdot \frac{du}{(1+u)^2}. \tag{99}$$

We consider ratio of the resonant cross section (96) to the conventional cross section of electron-nucleus bremsstrahlung in the case when an ultrarelativistic electron moves within the narrow cone with a photon producted in bremsstrahlung and scatters on the large angles. Using the results obtained by Baier et al. (1973) we may write the following expression:

$$R_{res} = \frac{d\sigma_{res}^{(1)}}{d\sigma_a} = \Upsilon\,(xy) \cdot \frac{E_i}{m\Gamma_\tau\sqrt{1+\mu_\tau^2}} \cdot \frac{dW^{(1)}}{dW_{p_i}\,(k')}. \tag{100}$$

Here, $dW_{p_i}(k')$ is the probability that an electron with the four-momentum p_i emits a photon with the four-momentum k'. For electron energies $m \ll E_i \ll m^2/\omega$ the expression (100) may be written as

$$R_{res} = \frac{\sqrt{a_2}}{8}\pi\frac{\eta_0^2\omega\tau}{\sqrt{1+\mu_\tau^2}} \cdot \frac{1}{\ln\left(E_i/m\right)}. \tag{101}$$

If the considered process characteristics satisfy the conditions (64), than the parameter $\mu_\tau \ll 1$ (70) and the resonant shape is specified by the laser pulse duration. Eq. (101) implies, when the ultrarelativistic electron energy grows, the resonant cross section decreases drastically.

3. Resonant photoproduction of an electron-positron pair on a nucleus in the field of a pulsed light wave

The most general computations of the resonant Coulomb electron-positron pair photoproduction (CPP) on a nucleus in the field of an electromagnetic plane wave was performed by Roshchupkin (1983). Borisov et al. (1981) studied the resonant CPP in the special case of ultrarelativistic electron and positron energies where the incident photon and the wave photon fly toward each other. The work of Lötstedt et al. (2008) in which resonant cross sections were calculated for strong external fields should also be noted. The resonant CPP in the pulsed light wave was studied in detail in the work of Lebed' & Roshchupkin (2011).

We consider the photoproduction of an electron-positron pair on a nucleus in a pulsed light field (2). The interaction of an electron and positron with a nucleus is considered in the first order of the perturbation theory (the Born approximation). Note that CPP is a crossed channel of bremsstrahlung due to electron scattering by a nucleus. Spontaneous bremsstrahlung of an electron scattered by a nucleus in a pulsed light field was studied early. In consideration of the known calculation procedure we may obtain the amplitude of CPP process on a nucleus in the field of a moderately strong pulsed wave from the expressions (9)-(18) by the following replacement:

$$p_- \to p_f, \quad p_+ \to -p_i, \quad k_i \to -k', \tag{102}$$

where p_-, p_+, k_i are the four-momenta of an electron, a positron and an initial photon, respectively. For CPP on a nucleus $q = (q_0, \mathbf{q})$ is the four-vector is the transferred momentum, q_- and q_+ are the four-momenta of an intermediate electron and an intermediate positron (for the diagrams on Fig. 7 (a) and (b), respectively). These quantities are expressed by the relationships:

$$\begin{cases} q = p_- + p_+ - k_i + lk, \\ q_- = k_i + rk - p_+, \\ q_+ = k_i + rk - p_-. \end{cases} \tag{103}$$

3.1 Resonance conditions

Let us consider the resonances that occur when an intermediate particle reaches the mass shell. The conditions of resonant CPP on a nucleus in a pulsed light field is determined by the relationship

$$q_\pm^2 - m^2 \lesssim \frac{(kq_\pm)}{\omega\tau}. \tag{104}$$

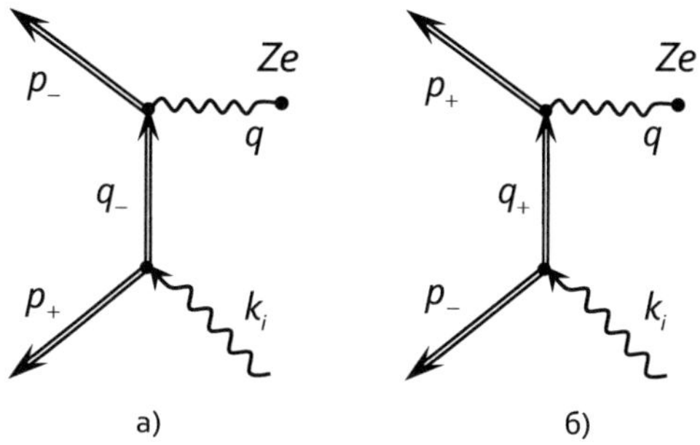

Fig. 7. Photoproduction of an electron–positron pair on a nucleus in a pulsed light wave.

Consequently, the four-momentum of an intermediate particle appears near the mass surface under the resonant conditions.

It is convenient to write Eqs. (103), which define the four-momenta q and $q_\pm$, for amplitudes (a) and (b) in Fig. 7, respectively, as

$$\begin{cases} k_i + rk = q_- + p_+, \\ q = p_- - q_- + (l + r)\, k; \end{cases} \tag{105}$$

$$\begin{cases} k_i + rk = p_- + q_+, \\ q = p_+ - q_+ + (l + r)\, k. \end{cases} \tag{106}$$

Eqs. (105)-(106) represent the four-momentum conservation laws for the diagrams vertices (Fig. 7) that, in view of the condition (104), hold only for $r > 0$.

Taking into account the condition (104) we will obtain the initial photon frequency ω_i^{res} for which a resonance can be observed (the resonant frequency) from the Eq. (105). Within the zeroth order with respect to the small parameter $(\omega\tau)^{-1}$ for the diagrams (a) and (b) (see Fig. 7), we obtain

$$\omega_i^{res} = \omega_i^{\pm} \equiv r\omega \cdot \frac{(np_\mp)}{(n_i q_\pm)}, \tag{107}$$

$$n = k/\omega = (1, \mathbf{n}), \quad n_i = k_i/\omega_i = (1, \mathbf{n}_i). \tag{108}$$

Within the region of moderately strong fields (5) the energy conservation law ($q_0 \approx 0$) may be written as

$$\omega_i \approx E_- + E_+. \tag{109}$$

Therefore, it follows from Eq. (107) that within the moderately strong fields region resonances are possible only for ultrarelativistic positron p_+ (diagram (a), Fig. 7) and electron p_- (diagram (b), Fig. 7), if they move within a narrow cone with the incident γ-ray photon k_i. In this case resonant frequencies (107) take the form

$$\omega_i^{\pm} = \frac{E_\pm}{1 - W_\pm/E_\pm}, \quad W_\pm = \frac{m^2}{r\omega} \cdot \frac{\left(1 + \delta_{i\pm}^2\right)}{2\,(nn_i)}, \tag{110}$$

where

$$\delta_{i\pm} = \theta_{i\pm} \cdot (E_\pm/m), \quad \theta_{i\pm} = \angle\,(\mathbf{k}_i, \mathbf{p}_\pm) \ll 1. \tag{111}$$

Hence the resonances are possible only for the electron (positron) energies above some threshold value $W_{\pm}$: $E_{\pm} > W_{\pm} \sim m^2/\omega$.

Using Eqs. (110) it is easy to obtain the positron energy at resonance:

$$E_{+} = \frac{1}{2}\left\{1 \pm \sqrt{1 - \frac{\omega_i^{th}}{\omega_i}}\right\} \cdot \omega_i, \tag{112}$$

where ω_i^{th} is the threshold frequency of an incident γ-ray photon,

$$\omega_i^{th} = \frac{2m^2}{\omega\left(1 - \cos\theta_i\right)}, \quad \theta_i = \angle\left(\mathbf{k}, \mathbf{k}_i\right). \tag{113}$$

As we see from Eq. (113), the threshold energy of an initial photon appreciably depends on its orientation relative to wave propagation direction. Thus, the threshold energy is minimal when an incident photon propagates towards the wave. In the opposite case, when an initial photon moves parallel to external field photons, no resonances are observed. Note that the electron energy can be obtained from Eq. (112) by reversing the sign in front of the square root. It follows from Eq. (112), that the energies of produced electron and positron near the threshold ($\omega_i - \omega_i^{th} \ll \omega_i^{th}$) are equal $E_{+} = E_{-} \cong \omega_i^{th}/2$. If, alternatively, the frequency of an incident γ-ray photon is great ($\omega_i \gg \omega_i^{th}$) then electron and positron energies differ considerably ($E_{+} = \omega_i - \omega_i^{th}/4 \approx \omega_i$, $E_{-} \approx \omega_i^{th}/4$).

The condition of interference of resonant amplitudes, that is $\omega_i^{+} = \omega_i^{-}$, assumes the form

$$\left(np_{-}\right)\left(n_i q_{-}\right) = \left(np_{+}\right)\left(n_i q_{+}\right). \tag{114}$$

Using the energy conservation law (109) and Eq. (110) we derive that the interference of resonant amplitudes appears when an electron p_{-} and a positron p_{+} propagate within a narrow cone with an incident γ-ray photon k_i, with $\delta_{i-} = \delta_{i+}$ and $\theta_{i-} \sim \omega/E_{-}$.

Below, we will consider the resonance of one diagram. We will assume that the initial photon frequency is

$$\omega_i \approx \omega_i^{res} = \omega_i^{-}. \tag{115}$$

3.2 Resonant amplitude

The amplitude of CPP on a nucleus in a pulsed light field under resonance conditions (107) has the form

$$S^{(\pm)} = \sum_{l=-\infty}^{\infty} S_l^{(\pm)}, \tag{116}$$

where $S_l^{(\pm)}$ is the partial amplitude, which corresponds to processes with emission ($l > 0$) or absorption ($l < 0$) of laser-wave $|l|$ photons

$$S_l^{(\pm)} = -i\frac{Ze^3\sqrt{\pi}}{\sqrt{2\omega_i E_{-} E_{+}}}\bar{u}_{-}\left[B_{l-}\left(\tilde{\gamma}_0, \hat{\varepsilon}_i\right) + B_{l+}\left(\hat{\varepsilon}_i, \tilde{\gamma}_0\right)\right]u_{+}. \tag{117}$$

Here, the functions $B_{l-}\left(\tilde{\gamma}_0, \hat{\varepsilon}_i\right)$ and $B_{l+}\left(\hat{\varepsilon}_i, \tilde{\gamma}_0\right)$ correspond to the CPP diagrams in Figs. 7(a) and 7(b), respectively

$$B_{l-}\left(\tilde{\gamma}_0, \hat{\varepsilon}_i\right) = \sum_{r=-\infty}^{\infty} \frac{2\omega\tau^2}{\mathbf{q}^2 + q_0\left(q_0 - 2q_z\right)} \cdot \frac{i\pi}{2\left(kq_-\right)} \times$$
$$\times \int_{-\infty}^{\infty} d\phi L_{l+r}\left(\phi\right) \exp\left\{i\left(q_0\tau + 2\beta\right)\phi\right\} \cdot \tilde{\gamma}_0\left(\hat{q}_- + m\right) \times \quad (118)$$
$$\times \int_{-\infty}^{\infty} d\phi' F_{-r}\left(\phi'\right) \exp\left\{-2i\beta\phi'\right\}\left(sgn\left(\phi' - \phi\right) - 1\right),$$

$$\beta = \frac{q_-^2 - m^2}{4\left(kq_-\right)}\omega\tau. \quad (119)$$

Here, functions $F_{-r}\left(\phi'\right)$ and $L_{l+r}\left(\phi\right)$ are defined by relations (15)-(21) with the replacement (102).

With allowance of the four-momentum conservation law (that is the first Eq. in (105)), the matrix function $F_{-r}\left(\chi_{q_-p_+}, \gamma_{q_-p_+}\left(\phi'\right), \beta_{q_-p_+}\left(\phi'\right)\right)$ (15) under resonance conditions defines the amplitude of the production of an electron-positron pair with the four-momenta q_- and p_+ by a photon with the four-momentum k_i in a pulsed light field through r wave photons absorption. This process was considered by Nikishov & Ritus (1979) in the case of a plane monochromatic wave, and by Narozhny & Fofanov (1997) in the case of a pulsed light wave. With allowance of the transferred four-momentum q (see the second equality in (105)) the quantity $L_{l+r}\left(\chi_{p_-q_-}, \gamma_{p_-q_-}\left(\phi\right), \beta_{p_-q_-}\left(\phi\right)\right)\tilde{\gamma}_0$ defines the amplitude of scattering of an intermediate electron with the four-momentum q_- by a nucleus in a pulsed light field with absorption or emission of $|l+r|$ photons of the wave (Lebed' & Roshchupkin (2008)).

Consequently, if the interference between direct and exchange amplitudes is absent, the process of resonant CPP on a nucleus in a pulsed light field effectively decomposes into two consecutive processes of the first order. The distinction for the diagram (b) on Fig. 7 is concluded in replacement of the four-momentum of an intermediate electron $q_- \rightarrow -q_+$ and change of sequence of first order processes.

Integral functions (18) are determined by the integer-order Bessel functions (23) for the case of a circularly polarized external wave. For circular polarization of a wave under resonance conditions the arguments of the Bessel functions (20) for CPP on a nucleus may be represented as

$$\gamma_{q_-p_+}\left(\phi'\right) = 2r \cdot \eta_0 g\left(\phi'\right) \cdot \frac{1+z_+}{z_+ z_r}\sqrt{z_+ z_r - \left(1+z_+\right)^2}, \quad (120)$$

where the invariant parameters z_+ and z_r are defined by

$$z_+ = \frac{\left(kp_+\right)}{\left(kq_-\right)} \approx \frac{E_+}{\omega_i - E_+}, \quad z_r = 2r \cdot \frac{\left(kk_i\right)}{m^2}. \quad (121)$$

It was expected for this part of the amplitude that the Bunkin-Fedorov quantum parameter becomes a classical one under resonance conditions (see Eqs. (32)-(33)).

$$\gamma_{q_-p_+}\left(\phi'\right) \sim \eta_0 \ll 1. \quad (122)$$

Consequently within the field range, specified by Eq. (5), the first resonance, that is, the resonance with $r = 1$, provides the main contribution to the resonant cross section, when the Bessel function has the largest value. This implies that the single-photon production of an electron–positron pair in a pulsed field proceeds mainly through absorption of one external field photon. However, the argument of the Bessel function $J_{l+r}\left(\gamma_{p_-q_-}(\phi)\right)$ is of the order of a magnitude $\gamma_{p_-q_-}(\varphi) \sim \gamma_0 \gtrsim 1$, i.e. it saves the quantum nature. Thus, scattering of an intermediate electron by a nucleus in the field of a moderately strong pulsed wave is a multiquantum process.

We perform the subsequent analysis for the case of wave circularly polarization ($\delta = \pm 1$) at expense of one wave photon absorption, i.e. $r = 1$. In view of the envelope function (40), after simple manipulations we obtain the amplitude (118) in the form

$$B_{l-}(\tilde{\gamma}_0, \hat{\varepsilon}_i) = \frac{2\pi \cdot \tilde{\gamma}_0 (\hat{q}_- + m)\hat{F}}{\mathbf{q}^2 + q_0(q_0 - 2q_z)} \cdot \frac{-i\omega\tau^2\sqrt{\pi}}{4(kq_-)} \exp\{-\frac{\beta^2}{4}\} \cdot I(q_0, \beta), \tag{123}$$

$$\hat{F} = -\frac{1}{2}\exp\{i\chi_{q_-p_+}\} \cdot \gamma_{q_-p_+}(0) \cdot \hat{\varepsilon}_i + \left((e_x + i\delta e_y)\, b\right), \tag{124}$$

$$I(q_0, \beta) = \int_{-\infty}^{\infty} d\phi \cdot J_{l+1}(\phi) \exp\{i(q_0\tau + 2\beta)\phi\} \left(\operatorname{erf}\left(2\phi + \frac{i\beta}{2}\right) + 1\right). \tag{125}$$

Here, $\operatorname{erf}(2\phi + i\beta/2)$ is the error function.

3.3 Resonant cross section

The differential cross section of CPP on a nucleus in a pulsed light field may be easily obtained by standard mode (Berestetskii et al. (1982)) from the amplitude, Eqs. (116)-(117), (123)-(125)

$$d\sigma^{(\pm)} = \sum_{l=-\infty}^{\infty} d\sigma_l^{(\pm)}, \tag{126}$$

where $d\sigma_l^{(\pm)}$ is the partial cross section of CPP on a nucleus in a pulsed light field with emission ($l > 0$) or absorption ($l < 0$) of $|l|$ wave photons.

Under resonance conditions and for ultrarelativistic electron and positron energies, the energy contribution from external pulsed field photons may be neglected. Therefore, the resonant cross section (126) may be summed over all possible partial processes. Thus, the differential cross section of CPP on a nucleus in a pulsed light field with the positron energy in the interval $[E_+, E_+ + dE_+]$ within the solid angle $[\Omega_+, \Omega_+ + d\Omega_+]$ and the final electron within the solid angle $[\Omega_-, \Omega_- + d\Omega_-]$ assumes the form

$$\frac{d\sigma_{1res}^{(\pm)}}{dE_+ d\Omega_+ d\Omega_-} = \frac{Z^2 e^6}{(2\pi)^2} \frac{|\mathbf{p}_-||\mathbf{p}_+|}{\omega_i \mathbf{q}^4} |\bar{u}_- M_- u_+|^2 \cdot P_{res}, \tag{127}$$

$$M_- = \tilde{\gamma}_0(\hat{q}_- + m)\hat{F}. \tag{128}$$

In Eq. (127) the function P_{res} is defined by the expression (53), where the replacement $q_i \to q_-$ has to be performed. We don't take polarization effects into consideration. After performing of corresponding averaging and summation procedures and considering that $d\Omega_{+} = \left(m^2/2E_{+}^2\right) d\delta_{i+}^2 d\varphi_{az}$ we derive

$$d\sigma_{1res}^{(\pm)} = \frac{1}{2\pi^2} \cdot \frac{m^2\omega_i}{z_+} \cdot P_{res} \cdot d\sigma_s\left(q_-\right) dW_{pair}^{(1)} d\delta_{i+}^2 d\varphi_{az}. \tag{129}$$

Here,

$$d\sigma_s\left(q_-\right) = 2Z^2 r_e^2 \frac{|\mathbf{p}_-|\, m^2}{|\mathbf{q}_-|\, \mathbf{q}^4} \left(m^2 + E_- q_{0-} + \mathbf{p}_- \mathbf{q}_-\right) d\Omega_- \tag{130}$$

is the differential cross section of scattering of an intermediate electron with the four-momentum q_- by a nucleus, and

$$dW_{pair}^{(1)} = \alpha \frac{\eta_0^2 m^2}{4\omega_i} \left\{ \frac{4\left(1+z_+\right)^2}{z_+ z_1} \left(1 - \frac{\left(1+z_+\right)^2}{z_+ z_1}\right) - 2 + \frac{\left(1+z_+\right)^2}{z_+} \right\} \cdot \frac{dz_+}{\left(1+z_+\right)^2} \tag{131}$$

is the probability of production of an electron-positron pair with the four-momenta q_- and p_+ by the an incident photon with the four-momentum k_i at the expense of one wave photon absorption. We can perform integration in Eq. (129) over the azimuthal angle $d\varphi_{az}$ and $d\delta_{i+}^2$. At that replacement $d\delta_{i+}^2 \to d\beta$ is to be carried out. The parameter β (119) under resonance conditions assumes the form

$$\beta = \frac{\omega\tau}{2} \left[1 - \frac{\left(1+z_+\right)^2}{z_+ z_1} \left(1 + \delta_{i+}^2\right)\right]. \tag{132}$$

We derive consequently

$$d\sigma_{1res}^{(\pm)} = \sqrt{\frac{\pi}{2}} \cdot \frac{\omega\tau}{2} \cdot \frac{\omega_i}{m^2 z_1} \cdot d\sigma_s\left(q_-\right) dW_{pair}^{(1)}. \tag{133}$$

Within the kinematical region of resonance, CPP on a nucleus in external field absence was investigated by Baier et al. (1973). It was concluded that amplitudes (a) and (b) (see Fig. 7) have poles within different regions of pair emission angles, therefore, they do not interfere. At that, the cross section is factorized, i.e.

$$d\sigma_{pair} = dW_{k_i}\left(p_+, q_-\right) \cdot d\sigma_S\left(q_-\right), \tag{134}$$

where $q_- = k_i - p_+$; $dW_{k_i}(p_+, q_-)$ is the probability of production of an electron-positron pair $(p_+ q_-)$ by an incident γ-ray photon with the four-momentum k_i. We express the resonant cross section (133) in terms of ordinary one (134),

$$R_{res} = \frac{d\sigma_{1res}^{(\pm)}}{d\sigma_{pair}} = \frac{\omega_i}{4m\Gamma_\tau\left(1+z_+\right)} \cdot \frac{dW_{pair}^{(1)}}{dW_{k_i}\left(p_+, q_-\right)}. \tag{135}$$

The transit width Γ_τ of the resonance was introduced here. It has the form

$$\Gamma_\tau = \sqrt{\frac{2}{\pi}} \cdot \frac{1}{\omega\tau} \cdot \frac{(kq_-)}{m}. \tag{136}$$

It is obvious from Eq. (136) that the transit width is specified by the pulsed field frequency and duration as well as by the particle energy and process kinematics. We underline that when CPP on a nucleus in the field of a plane monochromatic wave is studied the divergence in the differential cross section is eliminated by introducing of radiative corrections into the Green function of an intermediate particle according to the Breit-Wigner prescription as usual. It is concluded in addition of the imaginary part of the electron or positron mass: $m \to m - i\Gamma_R$. Here, the radiation width of resonance Γ_R is introduced phenomenologically. It has the form

$$\Gamma_R = \frac{1}{3}\alpha\eta_0^2 \cdot \frac{\sigma_c(q_-)}{\sigma_T} \cdot \frac{(kq_-)}{m}, \tag{137}$$

where $\sigma_c(q_-)$ is the total cross section of the Compton scattering of an external field photon by an intermediate electron with the four-momentum q_- (it is the most probable channel of an electron escape from an intermediate state), and σ_T is the Thompson cross section. Comparison of resonant widths (136) and (137) ascertains that the transit width exceeds the radiation one if laser pulse parameters satisfy the condition

$$\omega\tau < \frac{3}{\alpha\eta_0^2} \cdot \frac{\sigma_c(q_-)}{\sigma_T}. \tag{138}$$

Moderately strong fields of optical frequencies and the picosecond range of widths meet the inequality (138). The titanium-sapphire laser (Ti:Sapphire) or the solid-state laser based on aluminum-yttrium garnet $Y_2Al_5O_{12}$ with neodymium Nd admixtures (Nd:YAG) can be used as sources of such pulsed fields. Titanium-sapphire lasers have a broad lasing band (700-1100 nm) and a wide range of pulse duration (10 ps –10 fs) due to various choices of pulse compression. The PICAR picosecond Nd:YAG laser (designed at the International Educational-Scientific Laser Center of the Moscow State University named by M.V. Lomonosov) appropriate field characteristics to be achieved through the combined action of active-passive mode locking and a negative feedback (Gorbunkov et al. (2005)).

Ratio of cross-sections (135) is simplified considerably in the logarithmic approximation:

$$R_{res} = \frac{\pi}{8}\sqrt{\frac{\pi}{2}} \cdot \eta_0^2\omega\tau \cdot \left[\ln\frac{E_+}{m}\right]^{-1}. \tag{139}$$

Let us estimate the ratio of the cross sections (139) for PICAR picosecond Nd:YAG laser with additional amplifiers with parameters $\eta_0 \approx 0.1$, $\lambda = 1064$ nm ($\omega = 1.17$ eV), $\tau = 25$ ps. An incident γ-ray photon with an energy near the threshold value (113) $\omega_i = 5 \cdot 10^5 m = 255$ GeV propagates towards the pulsed laser wave. We obtain the following ratio of cross-sections: $R_{res} \approx 40$. Consequently, the resonant cross-section of CPP on a nucleus in a pulsed light field may exceed the corresponding one in external field absence by an order of magnitude.

4. Resonant scattering of a lepton by a lepton in the pulsed light field

Study of various processes of leptons scattering in an external electromagnetic fields is one of the fundamental directions of QED. Cross sections of basic scattering processes in the external field absence were obtained in the middle of the twentieth century. Thus, the scattering of an electron by an electron was considered by Möller (1932), the scattering of an electron by a positron - Bhabha (1938), the scattering of an electron by a muon - by Bhabha (1938) and Massey & Corben (1939). The detailed consideration of nonresonant scattering of an electron by a muon in a pulsed light field was performed by Padusenko et al. (2009).

We underline that the Bunkin–Fedorov quantum parameter γ_0 (4) is the main one which determines multiphoton processes in leptons nonresonant scattering. However in the case of leptons resonant scattering the influence of the quantum parameter γ_0 does not appear (it becomes a classical one due to resonance conditions and possess the values in order to η_0), thus the classical parameter η_0 (3) determines multiphoton processes. Therefore study of lepton by a lepton resonant scattering is carried out within the intensity range (5), that is within the framework of the first order of the perturbation theory with respect to an external laser field.

The electron mass m_e is considerably less than the muon one m_μ ($m_e \ll m_\mu$), therefore the corresponding classical parameters (3) satisfy the following condition as well

$$\eta_{0\mu} \ll \eta_{0e}. \tag{140}$$

The classical parameters $\eta_{0\mu}$ and η_{0e} are defined by Eqs. (3), where replacements $m \to m_\mu$ and $m \to m_e$ are to be performed. Hereinafter we consider resonances for direct Feynman diagrams of scattering type exceptionally (Fig. 8). Exchange diagrams for identical leptons and annihilation diagrams of scattering of a lepton by an antilepton are outside of attention. Such a problem statement is possible due to fact that resonances for direct diagrams of scattering type and resonances for exchange (annihilation) diagrams within the intensity range (5) occur within essentially different nonoverlapping kinematical regions (Roshchupkin & Voroshilo (2008)). For direct scattering amplitude within the fields range (5) the process of lepton by a lepton resonant scattering occurs when leptons scatter forwards into the small angles in the frame of the reference related to the center of inertia of initial particles and effectively decomposes into two processes of the first order similar to the Compton scattering of a wave by a lepton.

The S-matrix element for a direct amplitude (see Fig. 8) is given by

$$\begin{aligned} S = ie^2 \int d^4x_1 d^4x_2 D_{\mu\mu'}(x_1 - x_2) \times \\ \times \left[\bar{\psi}_{p_1'}(x_1|A)\tilde{\gamma}^{\mu}\psi_{p_1}(x_1|A)\right]\left[\bar{\psi}_{p_2'}(x_2|A)\tilde{\gamma}^{\mu'}\psi_{p_2}(x_2|A)\right]. \end{aligned} \tag{141}$$

Here, $D_{\mu\mu'}(x_1 - x_2)$ is the Green function of an intermediate free photon; $\psi_{p_j}(x|A)$ and $\bar{\psi}_{p_j'}(x|A)$ are the wave functions of initial and final leptons in the field of a pulsed light wave (2), respectively ($j = 1, 2$).

The amplitude of scattering of a lepton l_1 (with the mass m_1 and the four-momentum p_1) by a lepton l_2 (with the mass m_2 and the four-momentum p_2) in a pulsed light field may be represented as a sum of partial components with emission ($l > 0$) and absorption ($l < 0$) $|l|$ wave photons:

$$S = \sum_{l=-\infty}^{\infty} S_l, \tag{142}$$

$$S_l = \frac{(2\pi)^4 ie^2}{2\sqrt{E_1 E_2 E_1' E_2'}} \delta(q_x)\delta(q_y)\delta(q_0 - q_z) D_{ls}. \tag{143}$$

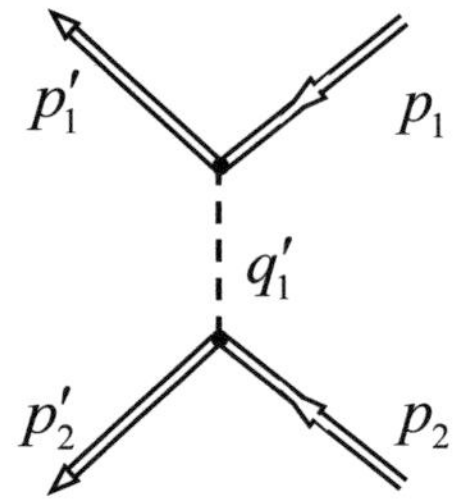

Fig. 8. The Feynman diagram of direct amplitude of scattering of a lepton l_1 by a lepton l_2 in the field of a pulsed light wave. External incoming and outgoing double lines correspond to the wave functions of leptons in initial and final states in the field of a plane wave (the Volkov functions), and an inner dashed line corresponds to a Green function of a free photon.

Here, the arguments of delta-functions are the four-vector $q = (q_0, \mathbf{q})$ components

$$q = p_1' + p_2' - p_1 - p_2 + lk. \tag{144}$$

The function D_{ls} in Eq. (143) has the form

$$D_{ls} = \sum_{s=-\infty}^{\infty} \int_{-\infty}^{\infty} \frac{\left(\bar{u}_{p_1'} \Lambda_{l-s}^{\nu}(\zeta) u_{p_1}\right)\left(\bar{u}_{p_2'} \Lambda_{s\nu}(\zeta) u_{p_2}\right)}{q_1'^2 + 2\zeta\left(kq'_1\right) + i0} d\zeta. \tag{145}$$

Here, q_1' is the four-vector of an intermediate photon

$$q_1' = p_2' - p_2 + sk = p_1 - p_1' + (l-s)k, \tag{146}$$

and functions $\Lambda_{l-s}^{\nu}(\zeta), \Lambda_{s\nu}(\zeta)$ are represented by

$$\begin{cases} \Lambda_{l-s}^{\nu}(\zeta) = \tau \int\limits_{-\infty}^{\infty} d\phi_1 \cdot G_{l-s}^{\nu}(\phi_1) \cdot \exp\left\{i\frac{q_0+q_z}{2}\tau\phi_1\right\} \cdot \exp\left\{-i\left(\zeta\omega\tau\right)\phi_1\right\}, \\ \Lambda_{s\nu}(\zeta) = \tau \int\limits_{-\infty}^{\infty} d\phi_2 \cdot G_{s\nu}(\phi_2) \cdot \exp\left\{i\left(\zeta\omega\tau\right)\phi_2\right\}. \end{cases} \tag{147}$$

Functions $G_{l-s}^{\nu}(\phi_1)$ in Eq. (147) have the form

$$\begin{aligned} G_{l-s}^{\nu}(\phi_1) &= a^{\nu} L_{l-s}(\phi_1) + \eta_{01}(\phi_1) \frac{m_1}{4\omega\kappa_1} \tilde{\gamma}^{\nu} \hat{k} \left[\hat{\varepsilon}_- L_{l-s+1}(\phi_1) + \hat{\varepsilon}_+ L_{l-s-1}(\phi_1)\right] + \\ &+ \eta_{01}(\phi_1) \frac{m_1}{4\omega\kappa_1'} \left[\hat{\varepsilon}_- L_{l-s+1}(\phi_1) + \hat{\varepsilon}_+ L_{l-s-1}(\phi_1)\right] \hat{k} \tilde{\gamma}^{\nu} + \\ &+ \left(1-\delta^2\right) \eta_{01}^2(\phi_1) \frac{m_1^2}{8\omega^2\kappa_1\kappa_1'} k^{\nu} \hat{k} \left[\left(L_{l-s+2}(\phi_1) + L_{l-s+2}(\phi_1)\right)\right], \end{aligned} \tag{148}$$

$$a^{\nu} = \tilde{\gamma}^{\nu} + \left(1+\delta^2\right) \eta_{01}^2(\phi_1) \frac{m_1^2}{4\omega^2\kappa_1\kappa_1'} k^{\nu} \hat{k}, \tag{149}$$

$$\eta_{0j}\left(\phi_j\right) = \eta_{0j} \cdot g\left(\phi_j\right). \tag{150}$$

Here, $\hat{\varepsilon}_{\pm}$ is the compression of four-vectors $\varepsilon_{\pm} = e_x \pm i\delta e_y$ with the Dirac's $\tilde{\gamma}^{\nu}$-matrices. The expression for the function $G_{s\nu}(\varphi_2)$ is ensued from Eqs. (148)-(149) by following indices replacement: $1 \to 2, l-s \to s, l-s\pm 1 \to s\pm 1, l-s\pm 2 \to s\pm 2$ and by the index ν omission also. By means of κ_j and κ'_j in functions $G^{\nu}_{l-s}(\phi_1)$ and $G_{s\nu}(\varphi_2)$ the following expressions are denoted

$$\begin{cases} \kappa_j = E_j - \mathbf{n}\mathbf{p}_j, \\ \kappa'_j = E'_j - \mathbf{n}\mathbf{p}'_j. \end{cases} \tag{151}$$

Here, $\mathbf{n}$ is the unit vector along the direction of external wave propagation

$$\mathbf{n} = \frac{\mathbf{k}}{|\mathbf{k}|}. \tag{152}$$

There are the integral functions $L_{l-s}(\phi_1)$, $L_s(\phi_2)$ which determine probability of emission and absorption of external wave photons in Eqs. for $G^{\nu}_{l-s}(\phi_1)$, $G_{s\nu}(\phi_2)$. They have the form

$$\begin{aligned} L_n\left(\phi_j\right) &\equiv L_n\left(\chi_j, \gamma_{0j}\left(\phi_j\right), \beta_j\left(\phi_j\right)\right) = \\ &= \frac{1}{2\pi}\int\limits_0^{2\pi} d\varphi'_j \exp\left\{i\left[\gamma_{0j}\left(\phi_j\right)\sin\left(\varphi'_j - \chi_j\right) + \beta_j\left(\phi_j\right)\sin 2\varphi'_j - n\varphi'_j\right]\right\} \end{aligned} \tag{153}$$

($j = 1$ for $n = l-s$, $j = 2$ for $n = s$) with the arguments

$$\gamma_{0j}\left(\phi_j\right) = \eta_{0j}\left(\phi_j\right)\cdot\frac{m_j}{\omega}\sqrt{\left(e_x g_j\right)^2 + \delta^2\left(e_y g_j\right)^2}, \tag{154}$$

$$tg\chi_j = \delta\frac{\left(e_y g_j\right)}{\left(e_x g_j\right)}, \quad g_j = \frac{p'_j}{\kappa'_j} - \frac{p_j}{\kappa_j}, \tag{155}$$

$$\beta_j\left(\phi_j\right) = \left(1-\delta^2\right)\eta_j^2\left(\phi_j\right)\frac{m_j^2}{8\omega}\left[\frac{1}{\kappa'_j} - \frac{1}{\kappa_j}\right]. \tag{156}$$

Before performing of integration of the function D_{ls} (145) over the variable ζ we remind that the subject of studying is the resonant character of amplitude behavior caused by quasi discrete structure: charged particle + plane electromagnetic wave. It is obvious that the resonant character of lepton-lepton scattering occurs when the denominator of the function D_{ls} approaches zero. We should underline that the possibility of lepton-lepton resonant scattering in a pulsed light field is provided by the both energy (with accuracy $q_0 \lesssim 1/\tau \ll \omega$) and momentum conservation laws fulfillment. Thus, the squared four-momentum of an intermediate photon q'_1 vanishes. It implies that the considered particle falls within the mass shell, i.e. an intermediate virtual photon becomes a real one. In this case the correction to the intermediate photon squared four-momentum in the denominator of the expression (145) is caused by the external field pulsed character and is essential through integration of the function D_{ls} (145) over the variable ζ. Hence, the following correlation is valid

$$q'^2_1 \lesssim \frac{(kq'_1)}{\omega\tau}. \tag{157}$$

The condition (157) determines such a kinematical region, which is accepted to name the resonant one. In the case of external field modeling as a plane monochromatic wave there is the intermediate particle squared four-momentum alone in a process amplitude denominator. Therefore, when a denominator is equal zero the resonant divergence occurs. It is eliminated by radiative corrections introducing into the Green function according the Breit–Wigner prescription. But now there is an addition in a denominator, caused by the laser wave pulsed character. Thus, the divergence in the process amplitude disappears.

Finally, the function D_{ls} (145) assumes the form:

$$D_{ls} = \sum_{s=-\infty}^{\infty} \frac{i\pi\omega\tau^2}{(kq_1')} \left(\bar{u}_{p'_\mu} \Delta^{\nu}_{l-s} u_{p_\mu}\right) \left(\bar{u}_{p'_e} \Delta_{s\nu} u_{p_e}\right), \tag{158}$$

where integral functions Δ^{ν}_{l-s}, $\Delta_{s\nu}$ are defined by following expressions

$$\begin{cases} \Delta^{\nu}_{l-s} = \int\limits_{-\infty}^{\infty} d\phi_1 \cdot G^{\nu}_{l-s}(\phi_1) \cdot \exp\left\{ i \left(\frac{q_0 + q_z}{2} \tau + 2\beta_\tau \right) \phi_1 \right\}, \\ \Delta_{s\nu} = \int\limits_{-\infty}^{\infty} d\phi_2 \cdot G_{s\nu}(\phi_2) \cdot \exp\{-2i\beta_\tau \phi_2\} \left(sgn(\phi_1 - \phi_2) - 1\right); \end{cases} \tag{159}$$

$$\beta_\tau \equiv \frac{q_1'^2}{4(kq_1')} \omega\tau. \tag{160}$$

Here β_τ is the relevant parameter which is defined by the both resonant scattering kinematics and external pulsed wave characteristics.

4.1 Resonance conditions

In this section we analyze in detail the case when an intermediate photon falls within the mass shell. Inner line discontinuity at the Feynman diagram appears and the studying process is effectively decomposes into two consecutive processes of the first order: a lepton l_1 with the four-momentum p_1 emits a real photon with the four-momentum q_1' at the expense of external wave photons absorption, then a real photon is absorbed by a lepton l_2 with external wave photons emission or vice versa.

Generally speaking owing to condition (157) the squared four-momentum of an intermediate photon is founded within the very narrow region near zero. We will show below that this region depends on initial four-momenta of scattered particles and their scattering angles. However, the given region has to be taken into consideration in the denominator of the resonant amplitude exceptionally (145). Thus, the four-momentum conservation laws for resonant diagram vertexes may be written as two equalities:

$$p_1 + |s|\,k = p_1' + q_1', \tag{161}$$

$$p_2 + q_1' = p_2' + s' \cdot k. \tag{162}$$

The equality (161) expresses the four-momentum conservation law in the process when an intermediate real photon is emitted by a lepton l_1 at the expense of $|s|$ external wave photons absorption. The equality (162) corresponds to the four-momentum conservation law in the process when an intermediate real photon is absorbed by a lepton l_2 with $s' = |s| + l$ external wave photons emission.

Remind that integral functions (153) are determined by the integer-order Bessel functions for the case of a circularly polarizated external wave. It is not difficult to verify that for this polarization under the resonance conditions (157) the arguments of the Bessel functions (154) may be represented as

$$\gamma_{0j}\left(\phi_j\right) = 2s' \cdot \eta_{0j}\left(\phi_j\right)\sqrt{\frac{u}{u_{s'}} \cdot \left(1 - \frac{u}{u_{s'}}\right)}. \tag{163}$$

$$u \equiv \frac{\left(kp_j\right)}{\left(kp'_j\right)} - 1 = \frac{\kappa_j}{\kappa'_j} - 1, \quad u_{s'} \equiv 2s' \cdot \frac{\omega\kappa_j}{m_j^2}. \tag{164}$$

It was expected, that for processes of resonant lepton-lepton scattering the influence of the Bunkin-Fedorov quantum parameter does not reveal, in opposite the nonresonant case. Since $\gamma_{0e} \sim \eta_{0e} \ll 1$ (see Eq. (163)), then the most probable case when a lepton l_1 absorbs and a lepton l_2 emits equally the only one external wave photon is realized, i.e.:

$$s' = |s| = 1, \quad l = s + s' = 0. \tag{165}$$

The region of resonant scattering is to be defined. We use the frame of reference related to a center of initial particles inertia, that is $\mathbf{p}_1 + \mathbf{p}_2 = 0$. In this frame the particle relative momentum $\mathbf{p} = \mathbf{p}_1 = -\mathbf{p}_2$ and after scattering changes only the direction: $|\mathbf{p}'| = |\mathbf{p}|$. We introduce also the unit vectors along the directions of initial and final momenta $\mathbf{n}_f$ and $\mathbf{n}_i$

$$\mathbf{n}_f = \frac{\mathbf{p}'}{|\mathbf{p}'|}, \quad \mathbf{n}_i = \frac{\mathbf{p}}{|\mathbf{p}|}. \tag{166}$$

With expressions (157) consideration it is easy to verify that in view of chosen direction of intermediate photon motion the resonance occurs if leptons scatter into the small angles in the frame of reference related to a center of inertia:

$$\theta = \angle\left(\mathbf{n}_f, \mathbf{n}_i\right) = \theta_{res} = 2\frac{\omega}{|\mathbf{p}|}\sin\theta_i \ll 1, \tag{167}$$

where $\theta_i = \angle\left(\mathbf{n}, \mathbf{n}_i\right)$ is the angle between the directions of wave propagation and the initial relative momentum p.

Meanwhile the resonance for exchange (annihilation) amplitude occurs in the essentially different kinematical region (see Roshchupkin & Voroshilo (2008)).

Thus, we expand the Bessel functions (148) as series in order of $\gamma_{0j} \sim \eta_{0j} \ll 1$ and keep the summands proportional to the first order of the parameter η_{0j}. Under the condition (165) we obtain:

$$G^{\nu}_{l-s}\left(\phi_1\right) = g\left(\phi_1\right) \cdot G^{\nu}_{1}, \tag{168}$$

$$G_{s\nu}\left(\phi_2\right) = g\left(\phi_2\right) \cdot G_{1\nu}. \tag{169}$$

where the matrices G_1^{ν} and $G_{1\nu}$ have the following form

$$G_1^{\nu} = (-1)\,\frac{\gamma_{01}}{2}\exp\left(i\chi_1\right)\tilde{\gamma}^{\nu} + \frac{\eta_{01}m_1}{2\omega\kappa_1}\left[k^{\nu}\hat{\varepsilon}_- - \varepsilon_-^{\nu}\hat{k}\right] + \frac{\eta_{01}m_1}{4\omega}\left(\frac{1}{\kappa_1'} - \frac{1}{\kappa_1}\right)\hat{\varepsilon}_-\hat{k}\tilde{\gamma}^{\nu}, \tag{170}$$

$$G_{1\nu} = \frac{\gamma_{02}}{2}\exp\left(-i\chi_2\right)\tilde{\gamma}_{\nu} + \frac{\eta_{02}m_2}{2\omega\kappa_2}\left[k_{\nu}\hat{\varepsilon}_+ - \varepsilon_{+\nu}\hat{k}\right] + \frac{\eta_{02}m_2}{4\omega}\left(\frac{1}{\kappa_2'} - \frac{1}{\kappa_2}\right)\hat{\varepsilon}_+\hat{k}\tilde{\gamma}_{\nu}. \tag{171}$$

The resonant region of scattering angles in the frame of reference related to a center of inertia is determined as

$$|\theta - \theta_{res}| \lesssim \frac{\theta_{res}}{\omega\tau} \ll \theta_{res}, \tag{172}$$

and expressions for the parameter β_τ (160) assumes the form

$$\beta_\tau = \frac{1}{2}\omega\tau\left(1 - \frac{\theta}{\theta_{res}}\right) \lesssim 1. \tag{173}$$

Finally, the resonant amplitude of a lepton l_1 scattered by a lepton l_2 in the field of a pulsed electromagnetic moderately strong wave of a circularly polarization in the frame of reference related to a center of inertia takes the form

$$S = S_0 \cdot \Upsilon_\tau, \tag{174}$$

where

$$S_0 = \frac{i\pi^{3/2}e^2\hat{M}}{\mathbf{p}^2\sqrt{E_1E_2E_1'E_2'}}\delta\left(P_x'\right)\delta\left(P_y'\right)\delta\left(E' - E - P_z'\right), \tag{175}$$

$$\hat{M} = \left(\bar{u}_{p_1'}G_1^{\nu}u_{p_1}\right)\left(\bar{u}_{p_2'}G_{1\nu}u_{p_2}\right). \tag{176}$$

The function Υ_τ in Eq. (174) is represented by

$$\Upsilon_\tau = \frac{\omega\tau}{\theta\cdot\theta_{res}}\exp\{-\frac{\beta_\tau^2}{4}\}\cdot I_\tau\left(q_+\tau\right). \tag{177}$$

Here, $I_\tau\left(q_+\right)$ is the integral function:

$$I_\tau\left(q_+\tau\right) = \tau\int_{-\infty}^{\infty} d\phi\cdot g\left(\phi\right)\cdot\exp\left\{i\left(\frac{q_0+q_z}{2}\tau + 2\beta_\tau\right)\phi\right\}\cdot\left[\operatorname{erf}\left(2\phi + \frac{i\beta_\tau}{2}\right) + 1\right]. \tag{178}$$

In Eqs. (177) and (178) the parameter β_τ is determined by the expression (173). We underline, that presence of three delta-functions in the resonant amplitude (174)-(178) is considered as realizing of three following conservation laws:

$$P_x' = 0, \quad P_y' = 0, \quad E' - E = P_z', \tag{179}$$

where $\mathbf{P}' = \left(P_x', P_y', P_z'\right)$ is the momentum of the inertia center after scattering, E and E' are particle total energies before and after scattering, correspondingly.

4.2 Resonant cross-section

In view of finite duration of an external pulsed light field there is a sense to define the differential probability over all the observation time T in the process of scattering of a lepton l_1 by a lepton l_2. Thus,

$$dW = |S|^2 \frac{d^3p'}{(2\pi)^3} \frac{d^3P'}{(2\pi)^3}. \tag{180}$$

Using the expressions for the amplitude (174)-(178) and performing uncomplicated computations we obtain the differential probability per time unit and per volume unit:

$$\begin{aligned} \frac{dW}{T} = dw = & \frac{e^4}{2(2\pi)^3 p^4 E_1 E_2 E_1' E_2' T} \cdot \frac{(\omega\tau)^2}{\theta^2 \theta_{res}^2} \cdot \left| \left(\bar{u}_{p_1'} G_1^{\nu} u_{p_1} \right) \left(\bar{u}_{p_2'} G_{1\nu} u_{p_2} \right) \right|^2 \times \\ & \times \exp\left(-\beta_\tau^2/2\right) \cdot |I_\tau (q_+\tau)|^2 \delta(P_x') \delta\left(P_y'\right) \delta\left(q_0 - P_z'\right) d^3p' d^3P'. \end{aligned} \tag{181}$$

The differential cross section we obtain from Eq. (181) by division by a density of the scattered particles flux $j = |\mathbf{p}|/E$. The integration of the differential cross section over d^3P' should be performed via the delta-functions. We present d^3p' as

$$d^3p' = E_1' E_2' |\mathbf{p}'| \, d\Omega' \frac{dE'}{E'}, \tag{182}$$

where $d\Omega'$ is the elementary solid angle of particles scattering, and introduce a new integration dimensionless variable: $dE' \to d\xi$ ($\xi = q_0/\omega$, $E' = \xi\omega + E$, $dE' = \omega d\xi$). After simple transformations we derive

$$\frac{d\sigma_{res}^{l_1 l_2}}{d\Omega'} = \frac{e^4 E}{2p^4 E_1 E_2} \cdot \frac{\omega(\omega\tau)^2}{\theta^2 \theta_{res}^2} \cdot \left| \left(\bar{u}_{p_1'} G_1^{\nu} u_{p_1} \right) \left(\bar{u}_{p_2'} G_{1\nu} u_{p_2} \right) \right|^2 \exp\left(-\beta_\tau^2/2\right) \cdot H. \tag{183}$$

Here, the function H has the form

$$\begin{aligned} H = \int_{-\infty}^{\infty} d\xi \frac{|I_\tau(\xi\omega\tau)|^2}{\xi\omega + E} = \tau^2 \int_{-\infty}^{\infty} \int_{-\infty}^{\infty} \int_{-\infty}^{\infty} \frac{d\xi d\phi d\phi'}{\xi\omega + E} g(\phi) \exp\left\{i(\xi\omega\tau + 2\beta_\tau)\phi\right\} \times \\ \times \left(\operatorname{erf}\left(2\phi + \frac{i\beta_\tau}{2}\right) + 1\right) g(\phi') \exp\left\{-i(\xi\omega\tau + 2\beta_\tau)\phi'\right\} \left(\operatorname{erf}\left(2\phi' - \frac{i\beta_\tau}{2}\right) + 1\right). \end{aligned} \tag{184}$$

The differential cross section of resonant scattering of nonpolarized leptons in the field of a pulsed light wave into the elementary solid angle may be represented as

$$\frac{d\sigma_{res}^{l_1 l_2}}{d\Omega'} = r_e^2 \frac{4\pi m_e^2 m_1^2 m_2^2}{p^4 E_1 E_2} \eta_{01}^2 \eta_{02}^2 \cdot f_0 \cdot f_{res}. \tag{185}$$

Here, the function f_0 is determined by

$$\begin{aligned} f_0 = & \left[\frac{2d_{fi} h_{fi} |\mathbf{p}| (E_1 + E_2)}{(E_1 - |\mathbf{p}| \cos\theta_i)(E_2 + |\mathbf{p}| \cos\theta_i)} \right]^2 + \left[2 + \frac{d_{fi}^2 (E_1 E_2 + \mathbf{p}^2)}{(E_1 - |\mathbf{p}| \cos\theta_i)(E_2 + |\mathbf{p}| \cos\theta_i)} \right] \times \\ & \times \left[2 + \frac{d_{fi}^2 (E_1 E_2 + \mathbf{p}^2) + 4d_{fi} h_{fi} |\mathbf{p}| (E_1 + E_2)}{(E_1 - |\mathbf{p}| \cos\theta_i)(E_2 + |\mathbf{p}| \cos\theta_i)} \right]. \end{aligned} \tag{186}$$

The following designations are used in Eq. (186)

$$h_{fi} = (\mathbf{e}_x \mathbf{n}_i) \cos \chi_{fi} + \delta \left(\mathbf{e}_y \mathbf{n}_i\right) \sin \chi_{fi}, \tag{187}$$

$$d_{fi} = 2 \left(\mathbf{n} \tau_{fi}\right) \sqrt{\left(\mathbf{e}_x \tau_{fi}\right)^2 + \left(\mathbf{e}_y \tau_{fi}\right)^2}, \tag{188}$$

$$tg\chi_{fi} = \delta \frac{\left(\mathbf{e}_y \cdot \tau_{fi}\right)}{\left(\mathbf{e}_x \cdot \tau_{fi}\right)}, \tag{189}$$

$$\tau_{fi} = \frac{\mathbf{n}_f - \mathbf{n}_i}{\left|\mathbf{n}_f - \mathbf{n}_i\right|}. \tag{190}$$

The function f_{res} in Eq. (185) has the form

$$f_{res} = \left(\frac{\omega\tau}{\theta_{res}^2}\right)^2 \cdot f\left(\rho, \beta_\tau\right), \tag{191}$$

$$f\left(\rho, \beta_\tau\right) = \exp\left(-\beta_\tau^2/2\right) \cdot \frac{1}{\rho} \int_{-\rho}^{\rho} d\phi g^2\left(\phi\right) \left|\mathrm{erf}\left(\phi + \frac{i\beta_\tau}{2}\right) + 1\right|^2. \tag{192}$$

We underline that the dependence of the function f_{res} on the parameter β_τ (173) determines resonant peak magnitude and shape. It is easy to notice that when leptons scatter into the resonant angle $\theta \simeq \theta_{res}$ than the parameter β_τ becomes equal zero (see Eq. (173)). At that the function f_{res} (191) possesses the finite value as opposed to the plane monochromatic wave case when $f_{res} \to \infty$ is correct.

The significant issue is the influence of the pulse finite duration on the cross section resonant behavior. The pulse duration has to exceed the time required for the Compton scattering of an external field photon by each of leptons l_1 and l_2. If this condition is not satisfied than particles do not have time to interact with a wave under the resonance conditions. Consequently, the following correlation for the pulse duration is valid:

$$\omega\tau \gtrsim \frac{1}{\alpha\eta_{0j}^2} \frac{E_j}{\kappa_j}. \tag{193}$$

Thus, experimental treatment of resonant scattering of a lepton by a lepton may be verified in the fields created by picosecond pulsed lasers which generate the radiation within the optical frequencies range. Such scientific facilities are employed in SLAC National Accelerator Laboratory (Bula et al. (1996); Burke et al. (1997)) research centers and also in the frame of the FAIR project (Bagnoud et al. (2009)).

We can integrate the differential cross section (185) within the narrow range of scattering angles near the resonance (172). Under the resonance conditions the vector τ_{fi} (190) may be represented as

$$\tau_{fi} \approx \frac{1}{\theta_{\mathrm{res}}} \left(\mathbf{n}_f - \mathbf{n}_i\right), \tag{194}$$

and $\left(\mathbf{n}_i \tau_{fi}\right) \simeq 0$. We perform the integration over the parameter β_T (173) (instead the scattering angle θ), and finally derive

$$\frac{d\sigma_{res}^{l_1 l_2}}{d\varphi_f} = 16\pi r_e^2 \eta_{01}^2 \eta_{02}^2 \frac{m_e^2 m_1^2 m_2^2}{E_1 E_2 \left|p\right|^4} \frac{(\omega\tau)}{\theta_{res}^2} \cdot f_0 \cdot F(\rho), \tag{195}$$

where function $F(\rho)$ is determined by

$$F(\rho) = \int_{-\infty}^{\infty} d\beta_T \cdot f(\rho, \beta_T). \tag{196}$$

Here, the function $f(\rho, \beta_T)$ is specified by Eq. (192). The limits of the integration in Eq. (196) are extended over the infinity owing to the integral quick convergence (though the values of the parameter $\beta_T \lesssim 1$ within the resonant region).

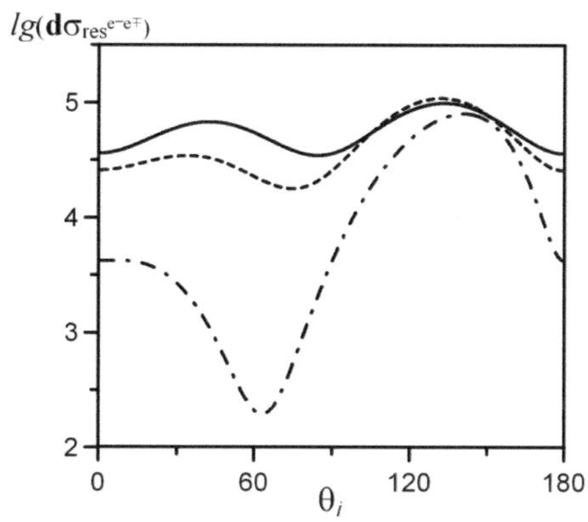

Fig. 9. The dependence or the differential cross-section of scattering of an electron by an electron (an electron by a positron) in a pulsed light field (195) (in units of respective cross-sections in an external field absence) on the initial polar angle when an azimuthal angle is fixed $\varphi_i = \pi/4$ and value of the parameter $\rho = 2$. The external laser wave frequency amounts to the value $\omega = 2.35$ eV, the pulse duration is equal to $\tau = 1.5$ ps, the field strength in a pulse peak $F_0 = 6 \cdot 10^9$ V/cm. The cases of particles relative velocities $V = 0.2$ (solid line), $V = 0.6$ (dotted line), and $V = 0.9$ (dash-dotted line) are represented.

Let us consider the ratio of the derived resonant differential cross section (195) to the differential cross section of scattering of the same leptons in an external field absence for such processes: scattering of an electron by an electron, scattering of an electron by a positron, scattering of an electron by a muon. Figs. 9, 10 show the dependencies of the considered ratio on the initial polar angle θ_i. We should underline that under scattering of both an electron by an electron and an electron by a positron within the small angles range (172) the respective cross-sections coincide each with other.

In accordance with the Figs. 9, 10 we consider that within the broad range of particles velocities the resonant cross sections of scattering of an electron by an electron (an electron by a positron, an electron by a muon) in a pulsed light field exceed the corresponding differential

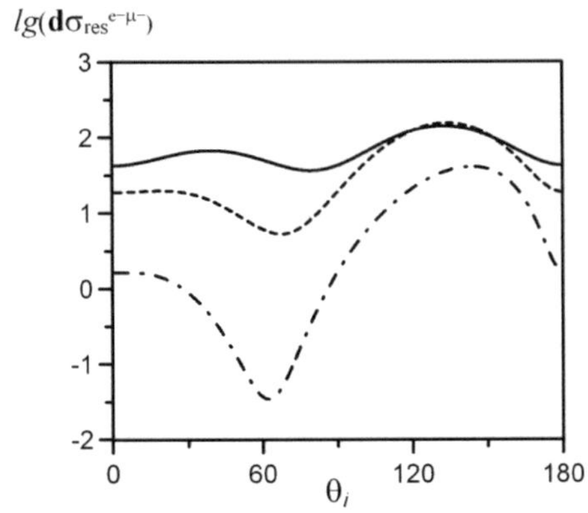

Fig. 10. The dependence of the differential cross-section of scattering of an electron by muon in a pulsed light field (195) (in units of respective cross-sections in an external field absence) on the initial polar angle when an azimuthal angle is fixed $\varphi_i = \pi/4$ and the value of the parameter $\rho = 2$. The cases of particles relative velocities $V = 0.2$ (solid line), $V = 0.6$ (dotted line) and $V = 0.9$ (dash-dotted Line) are represented.

cross sections in an external field absence within the whole polar angles range. Hereby, the greatest exceeding appears for the case of particles small relative velocities ($V = 0.2$), at that the exceeding reaches into five orders of the magnitude (for scattering of an electron by an electron (positron)), and two orders for scattering of an electron by a muon. Also there is a suppression of the resonant cross section in the case of leptons high relative velocities within the range of the initial polar angles $\theta_i \approx 60^\circ$.

5. Resonant scattering of a photon by an electron in the pulsed laser field

Oleinik (1967) specified resonances in the Compton effect in the field of a plane monochromatic wave for the first time, but his studies had a rather fragmentary form (see also Belousov (1977)). The resonance of direct and exchange diagrams in the general relativistic case for the field of a weakly intensive plane monochromatic electromagnetic wave was considered by Voroshilo & Roshchupkin (2005). Scattering of a photon by an electron in a pulsed light field for the direct diagram resonance in the range of weak fields (5) was studied in work Voroshilo et al. (2011).

5.1 Process amplitude

The amplitude of scattering of a photon with the four momentum $k_i = (\omega_i, \mathbf{k}_i)$ by an electron with the four momentum $p_i = (E_i, \mathbf{p}_i)$ in an pulsed field (2) (Fig. 11) is given by the expression

$$S_{fi} = S_{fi}^{(d)} + S_{fi}^{(e)}, \tag{197}$$

$$S_{fi}^{(d)} = -ie^2 \int d^4r d^4r' \bar{\Psi}_{p_f}(r) \tilde{\gamma}^\mu G(r, r') \tilde{\gamma}^\nu \Psi_{p_i}(r') A_\mu^*(k_f r) A_\nu(k_i r'), \quad S_{fi}^{(e)} = S_{fi}^{(d)}\left(k_f \leftrightarrow -k_i\right), \tag{198}$$

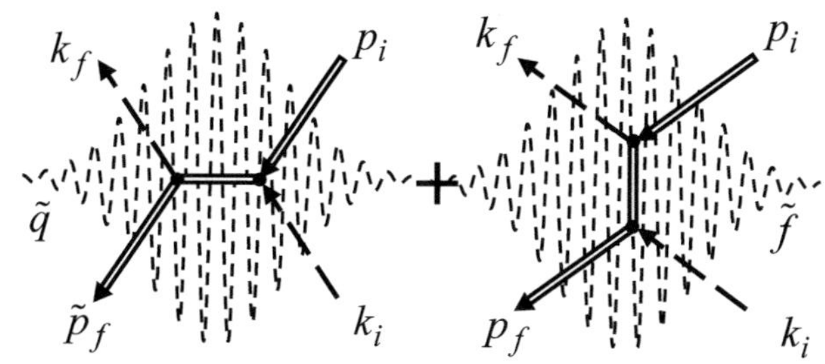

Fig. 11. The Feynman diagram of the Compton effect in the field of a pulsed light wave.

where $p_f = (E_f, \mathbf{p}_f)$ and $k_f = (\omega_f, \mathbf{k}_f)$ are four momenta of an outgoing electron and a photon; $\tilde{\gamma}^\nu$ ($\nu = 0, 1, 2, 3$) are the Dirac matrices; $A_\mu(k_i r')$ is the wave function of a photon (8); e_μ is the photon polarization four-vector; $G(r, r')$ is the Green function of an electron in the field (2).

The case when a laser field intensity meets the following condition

$$\eta_0^2 \lesssim \varphi_0^{-1} \ll 1, \quad \varphi_0 = \omega\tau, \tag{199}$$

is considered through this section. This condition allows both to carry out the decomposition with respect to the small parameter and to neglect the interference of contributions of the pulsed wave anterior and posticous parts.

The amplitude (197) accurate within terms $\sim \eta_0^2$ assumes the form

$$S_{fi} \approx B\delta^{(2)}(\mathbf{p}_{i,\perp} + \mathbf{k}_{i,\perp} - \mathbf{p}_{f,\perp} - \mathbf{k}_{f,\perp})\delta(p_{i,-} + k_{i,-} - p_{f,-} - k_{f,-})e'^*_\nu e_\mu \cdot \bar{u}_{p_f} T^{\nu\mu}_{fi} u_{p_i}, \tag{200}$$

$$T^{\nu\mu}_{fi} = \sum_j \left(T^{(j)\nu\mu}_{0,0} + \eta_0 \sum_{l,l' \in (|l-l'|+|l'|=1)} T^{(j)\nu\mu}_{l-l',l'} + \eta_0^2 \left(T^{(j)\nu\mu}_{0,0} + \sum_{l,l' \in (|l-l'|+|l'|=2)} T^{(j)\nu\mu}_{l-l',l'} \right) \right), \tag{201}$$

where $j = e, d$; indices d, e are concerned to direct and exchange diagrams; B is the normalization factor; $\mathbf{p}_{i,\perp}$, $\mathbf{k}_{i,\perp}$, $\mathbf{p}_{f,\perp}$, $\mathbf{k}_{f,\perp}$ are the projections of corresponding vectors on the wave polarization plane; $p_{i,-} = E_i - p_{i,z}$, $k_{i,-} = \omega_i - k_{i,z}$, $p_{f,-} = E_f - p_{f,z}$, $k_{f,-} = \omega_f - k_{f,z}$ are differences between zeroth components of the corresponding four momentum and its projection on direction of wave propagation; q, f are four momenta of an intermediate particle, which conform to direct and exchange diagrams on Fig. 11, at that under the four momenta conservation laws we have

$$\mathbf{q}_\perp = \mathbf{p}_{i,\perp} + \mathbf{k}_{i,\perp}, \quad q_- = p_{i,-} + k_{i,-}; \quad \mathbf{f}_\perp = \mathbf{p}_{i,\perp} - \mathbf{k}_{f,\perp}, \quad f_- = p_{i,-} - k_{f,-}. \tag{202}$$

The summands in Eq. (201), proportional to the zeroth degree of η_0 determine the amplitude of the Compton effect in external field absence (Klein & Nishina (1929)). The summands, proportional to the first degree of the parameter η_0, determine the corrections (for them $|l - l'| + |l'| = 1$ is valid) specified by participation of one wave photon in the process. The summands, proportional to the second degree of the parameter η_0, determine the corrections specified by participation of two wave photons in the process (for them $|l - l'| + |l'| = 2$ is valid).

In the case of a plane monochromatic wave the resonance is associated with the fact that an intermediate particle falls within the mass shell: $q^2 = m^2$, $f^2 = m^2$. The corrections to the Compton effect probability, which are specified by processes with one wave photon participation, are the nonresonant. They are proportional to the second order of the parameter η_0 and, therefore, are small in comparison with the Compton effect probability. But among processes with two wave photons participation there are such ones, which may have the resonant behavior. The both resonance of direct diagram through an electron intermediate state and resonance of exchange diagram through a positron intermediate state permit the processes with $l' = -1$, $l = 0$. The resonance of the exchange diagram through an electron intermediate state permits the process with $l' = 1$, $l = 0$. These processes may have resonant character in the case of a pulsed field (2) (Voroshilo et al. (2011)).

The expressions for $T^{(d,e)\nu\mu}_{l-l',l'}$ in Eq. (201) for resonant processes have the form:

$$T^{(d)\nu\mu}_{1,-1} \approx \frac{\pi\omega}{8(kq_{-1})} \cdot I(\beta_{-1}(q_{-1}), l_*) \left[M^{\nu}_{1}(p_f, q_{-1}) \left(\hat{q}_{-1} + m\right) M^{\mu}_{-1}(q_{-1}, p_i) \right], \tag{203}$$

$$T^{(e)\nu\mu}_{\mp 1,\pm 1} \approx \frac{\pi\omega}{8(kf_{\mp 1})} \cdot I(\beta_{\mp 1}(f_{\mp 1}), l_*) \left[M^{\nu}_{\pm 1}(p_f, f_{\mp 1}) \left(\hat{f}_{\mp 1} + m\right) M^{\mu}_{\mp 1}(f_{\mp 1}, p_i) \right]. \tag{204}$$

Here

$$I(\beta_{l'}, l_*) = \frac{\pi}{4(kq_{l'})} \left(\mathrm{erfi}\left(\frac{\sqrt{2}}{2} \left(\beta_{l'} - \frac{l_*\varphi_0}{4} \right) \right) + i \right) \exp\left\{ -\frac{\varphi_0^2 l_*^2 + 8\left(\beta_{l'} - l_*\varphi_0/4\right)^2}{16} \right\}, \tag{205}$$

where erfi(z) is the error function of imaginary argument; $\beta_{l'}$ is the resonant parameter:

$$\beta_{l'}\left(q_{l'}\right) = \frac{q_{l'}^2 - m^2}{4(kq_{l'})} \omega\tau. \tag{206}$$

Exactly the parameter $\beta_{l'}$ determines the process behavior character. Thus, the values $\beta_{l'} \lesssim 1$ correspond to the resonant behavior. The opposite case $\beta_{l'} \gg 1$ corresponds to the nonresonant one. Under the values $\beta_{l'} \gg 1$ the function $I(\beta_{l'}, a)$ has the following asymptotic form:

$$I(\beta_{l'}, l_*) \approx \sqrt{\frac{\pi}{2}} \frac{1}{\beta_{l'}} \exp\left\{ -\frac{1}{32} l_*^2 \varphi_0^2 \right\}. \tag{207}$$

In Eqs. (203)-(206) the quantities $q_{-1} = p_i + k_i - k$, $f_{\mp 1} = p_i - k_f \pm k$ correspond to the "strict" four momentum conservation law (like the monochromatic wave case, when summands $\sim \eta_0^2$ are neglected); the quantity l_* are the invariant parameter which are determined from the following equation:

$$p_i + k_i + l_* k = p_f + k_f. \tag{208}$$

It follows from Eq. (205) that $|l_*| \sim \varphi_0^{-1}$. Consequently, in the zero-order approximation with respect to the parameter φ_0^{-1} the frequency of a scattered photon is amount:

$$\omega_f \approx \omega_f^{(0)}, \quad \omega_f^{(0)} = \frac{(p_i k_i)}{E_i + \omega_i - \left([\mathbf{p}_i + \mathbf{k}_i] \mathbf{n}_f \right)}, \tag{209}$$

where $\mathbf{n}_f$ is the directive unit vector for the final photon emission.

Bispinor matrices $M^{\nu}_{\pm 1}$ in (203), (204) are determined by:

$$M^{\nu}_{\pm 1}(p_f, q_{-1}) = \pm \frac{y_0(p_f, q_{-1})}{2} e^{\mp i\chi} \cdot \gamma^{\nu} + \\ + \frac{m}{4}\left(\frac{2}{(kq_{-1})}\left[\hat{\varepsilon}^{(\mp)}k^{\nu} - \hat{k}\varepsilon^{(\mp)\nu}\right] + \left[\frac{1}{(kp_f)} - \frac{1}{(kq_{-1})}\right]\hat{\varepsilon}^{(\mp)}\hat{k}\gamma^{\nu}\right). \tag{210}$$

Here, quantities $\varepsilon^{(\pm)} = e_x \pm i\delta e_y$; $y_0(p_f, q), \chi \equiv \chi(p_f, q)$ are the kinematical parameters

$$y_0(p_f, q) = m\eta\sqrt{-g^2(p_f, q)}, \quad \tan\chi = \frac{\delta(ge_y)}{(ge_x)}, \quad g \equiv g(p_f, q) = \frac{p_f}{(kp_f)} - \frac{q}{(kq)}. \tag{211}$$

5.2 Resonant kinematics

5.2.1 Resonance conditions for the direct diagram

The parameter $\beta \equiv \beta(q_{-1})$ which corresponds to the resonant process with $l' = -1$, $l = 0$ (one field photon emits in the beginning, and one photon absorbs at the end of the process) may be written in the form:

$$\frac{\beta}{\varphi_0} = \frac{1}{2}\frac{1-\tilde{u}}{\left[1 + \tilde{u}\left(\omega_i/\omega_{i,\text{res}} - 1\right)\right]}\left(\frac{\omega_i}{\omega_{i,\text{res}}} - 1\right). \tag{212}$$

Here, the invariant parameter $\tilde{u}$ and the frequency $\omega_{i,\text{res}}$, which corresponds to the resonant maximum, are determined by:

$$\tilde{u} = \frac{(kk_i)}{(p_i k_i)}, \quad 0 \le \tilde{u} \le u_1, \quad u_1 = \frac{2(kp_i)}{m^2}, \tag{213}$$

$$\omega_{i,\text{res}} = \frac{(kp_i)}{E_i - \omega - ([\mathbf{p}_i - \mathbf{k}]\mathbf{n}_i)}, \tag{214}$$

where $\mathbf{n}_i$ is the unit vector along the propagation direction of incident photon. We rewrite this expression as

$$\omega_{i,\text{res}} = \frac{mu_1}{2}\frac{(E_i - \omega)/m + \sqrt{(E_i - \omega)^2/m^2 + u_1 - 1}\cdot\cos\tilde{\theta}_{\mathbf{S}}}{1 - u_1 + \left((E_i - \omega)^2/m^2 + u_1 - 1\right)\sin^2\tilde{\theta}_{\mathbf{S}}}, \tag{215}$$

where

$$\tilde{\theta}_{\mathbf{S}} = \angle(\mathbf{S}, \mathbf{n}_i), \quad \mathbf{S} = \mathbf{p}_i - \mathbf{k}. \tag{216}$$

We consider that the correlation $\omega \ll m$ is valid in the region $v_i = |\mathbf{p}_i|/E_i \ll \omega/m \ll 1$ (it is the nonrelativistic case, which also corresponds to the rest frame of an final electron) and obtain:

$$\omega_{i,\text{res}} = \frac{\omega}{1 - \omega/m\,(1 - \cos\tilde{\theta})} \approx \omega\left(1 + \frac{\omega}{m}(1 - \cos\tilde{\theta})\right), \quad \tilde{\theta} = \angle(\mathbf{k}, \mathbf{n}_i) \approx \pi - \tilde{\theta}_{\mathbf{S}}. \tag{217}$$

Therefore, in this case the resonant frequency is closely approximated to the laser field frequency.

In range where the correlation $\omega/m \ll v_i < 1$ is valid (it is the ultrarelativistic case) we derive:

$$\omega_{i,\mathrm{res}} \approx \frac{mu_1}{2} \frac{E_i/m + \sqrt{E_i^2/m^2 - 1}\cos\tilde{\theta}_{\mathbf{S}}}{1 - u_1 + \left(E_i^2/m^2 - 1\right)\sin^2\tilde{\theta}_{\mathbf{S}}}.$$

In the ultrarelativistic case ($u_1 > 1$, $E_i/m > m/\omega \gg 1$) under $(m/\omega)\sqrt{u_1 - 1} < \tilde{\theta}_{\mathbf{S}} \ll 1$ ($\tilde{\theta}_{\mathbf{S}} \approx \tilde{\theta}_{\mathbf{p}_\mathrm{i}} = \angle(\mathbf{p}_i, \mathbf{n}_i)$) we obtain:

$$\omega_{i,\mathrm{res}} \approx \frac{u_1 E_i}{1 - u_1 + (E_i/m)^2\,\tilde{\theta}_{\mathbf{S}}^2}. \tag{218}$$

Fig. 12 demonstrated dependence of the resonant frequency on the angle $\tilde{\theta}_{\mathbf{S}}$ for different energies of an electron.

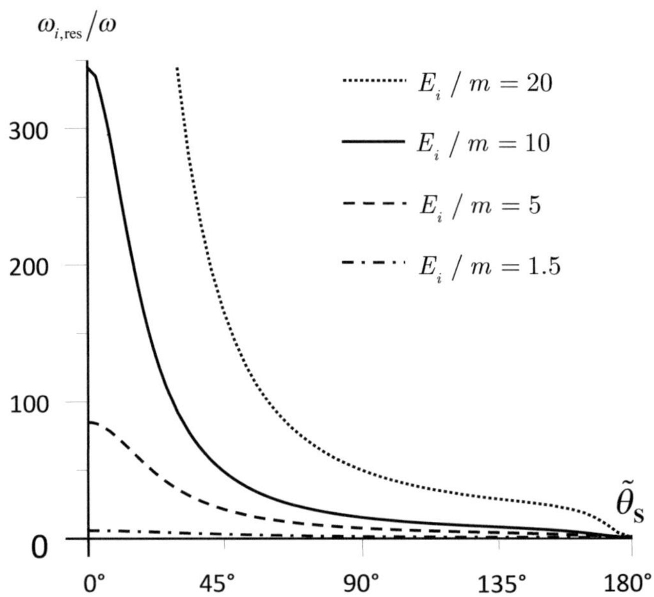

Fig. 12. The dependence of ratio of the resonant frequency of an ingoing photon to the laser field frequency $\omega_{i,\mathrm{res}}/\omega$ (215) from the angle $\tilde{\theta}_{\mathbf{S}}$ (216) under $\omega/m = 10^{-5}$ for different energies of an ingoing electron.

The resonance of the amplitude, which corresponds to the direct diagram, is feasible only when the condition $\tilde{u} < 1$ is satisfied, so that for the values $u_1 > 1$ the angle $\tilde{\theta}_{\mathbf{S}}$ is restricted by the interval:

$$\alpha_0 < \tilde{\theta}_{\mathbf{S}} < \pi, \quad \alpha_0 = \arccos\frac{E_i - \omega}{|\mathbf{S}|}. \tag{219}$$

The cases close to realization of the condition $\tilde{u} = 1$ $(\tilde{\theta}_{\mathbf{S}} = \alpha_0)$ also have to be excluded, because the frequency of a resonant photon in these cases has to be infinite, but it is impossible to put into practice. Therefore, the condition of the direct diagram resonance is determined by:

$$\left|1 - \frac{\omega_i}{\omega_{i,\mathrm{res}}}\right| \sim \frac{1}{\varphi_0} \ll 1. \tag{220}$$

Under the condition (220) the resonant parameter assumes the form:

$$\frac{\beta}{\varphi_0} \approx \frac{1}{2}(1-\tilde{u})\left(\frac{\omega_i}{\omega_{i,\mathrm{res}}} - 1\right), \quad 0 < \tilde{u} < \begin{cases} u_1, & u_1 < 1; \\ 1, & u_1 > 1. \end{cases} \tag{221}$$

5.2.2 Resonance conditions for the exchange diagram

For the processes with $l' = \pm 1$, $l = 0$ which permit the resonance of the exchange diagram through an electron $(l' = 1)$ and a positron $(l' = -1)$ intermediate states the resonant parameters $\beta_{\mp} \equiv \beta(f_{\pm 1})$ have the form:

$$\frac{\beta_{\mp}}{\varphi_0} = \frac{1}{2}\frac{v' \pm 1}{\left[v'\left(1 - \omega_f/\omega^{(\mp)}_{f,\mathrm{res}}\right) \pm 1\right]}\left(1 - \frac{\omega_f}{\omega^{(\mp)}_{f,\mathrm{res}}}\right), \tag{222}$$

where the upper sign is concerned to an electron intermediate state, the lower sign is concerned to a positron one; the invariant parameter v' and the frequencies $\omega^{(\mp)}_{f,\mathrm{res}}$ of a final photon, which correspond to the resonant maximum, are defined by:

$$v' = \frac{\left(kk_f\right)}{\left(p_i k_f\right)}, \quad \omega^{(\mp)}_{f,\mathrm{res}} = \frac{(kp_i)}{(E_i - \mathbf{p}_i\mathbf{n}_f)(v' \pm 1)}. \tag{223}$$

It follows from Eq. (223) that the resonance via positron intermediate state can be observed under limitations on parameter v' and, hence, angle $\tilde{\theta}'_{\mathbf{S}} = \angle(\mathbf{S}, \mathbf{n}_f)$:

$$v' > 1 (u_1 > 1) \quad 0 \le \tilde{\theta}'_{\mathbf{S}} \le \alpha_0 \quad \text{and} \quad \pi - \alpha_0 \le \tilde{\theta}'_{\mathbf{S}} \le \pi. \tag{224}$$

Equating the expressions (209), (223) we obtain that under the exchange diagram resonance directions of a scattered photon correspond to the condition of the resonant maximum; these directions lie on the surface of a cone (see Fig. 13); axis of the cone coincides with the vector $\mathbf{h}^{\mp}$ and the opening angle $\theta'_{\mathbf{h}^{\mp}} = \angle(\mathbf{h}^{\mp}, \mathbf{n}_f)$:

$$\cos\theta'_{\mathbf{h}^{\mp}} = h^{\mp}_0/|\mathbf{h}^{\mp}|, \quad h^{\mp} = (h^{\mp}_0, \mathbf{h}^{\mp}) = (kp_i)[p_i + k_i] - (p_i k_i)[k \pm p_i]. \tag{225}$$

Thus, the four vector has to be a spatially similar one $(h^{\mp})^2 \le 0$, i.e. the inequality

$$\tilde{u}_1^2(1 - \tilde{u}u_1) \mp 2\tilde{u}_1 u_1 + u_1^2 \le 0, \tag{226}$$

has to be met.

The invariant parameter $\tilde{u}_1$ is equal to:

$$\tilde{u}_1 = \frac{2(pk_i)}{m^2}. \tag{227}$$

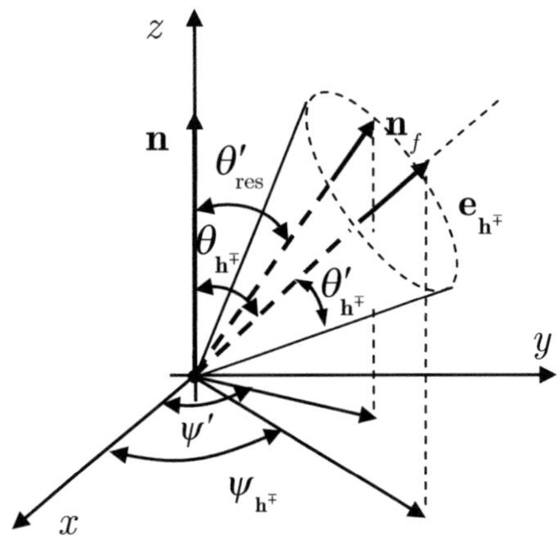

Fig. 13. Geometry of emission of an outgoing photon in the case of occurrence of the exchange diagram resonance.

From the inequality (226) we derive the following condition on the initial photon frequency; at this frequency the exchange diagram resonance through the electron intermediate state occurs:

$$\begin{cases} \dfrac{\omega_f}{1+\sqrt{u_1\tilde{u}}} \leq \omega_{i,\mathrm{res}} \leq \dfrac{\omega_f}{1-\sqrt{u_1\tilde{u}}}, & \tilde{u} < u_1^{-1}; \\ \omega_{i,\mathrm{res}} \geq \dfrac{\omega_f}{1+\sqrt{u_1\tilde{u}}}, & u_1^{-1} < \tilde{u} < u_1. \end{cases} \tag{228}$$

Here, the function f has the form

$$f = \frac{1 - v_i \cos\theta}{1 - v_i \cos\tilde{\theta}}, \tag{229}$$

where $\theta = \angle(\mathbf{k}, \mathbf{k}_i)$, $\tilde{\theta} = \angle(\mathbf{k}, \mathbf{p}_i)$.

For a positron intermediate state the resonance occur under the condition that the initial photon frequency exceeds a certain threshold value:

$$\omega_{i,\mathrm{res}} \geq \frac{\omega_f}{\sqrt{u_1\tilde{u}} - 1}, \quad u_1^{-1} < \tilde{u} < u_1. \tag{230}$$

Values of initial photon frequencies meet the condition of the direct diagram resonance $\omega_i = \omega_{i,\mathrm{res}}$ (214). They are founded within the frequencies interval (228); the exchange diagram resonance through an electron intermediate state occurs under these frequencies. Consequently, the direct diagram resonance is always accompanied by the exchange diagram resonance through an electron intermediate state, and within the region

$$u_1 > 1, \quad 1/u_1 < \tilde{u} < 1 \tag{231}$$

through a positron intermediate state also.

Fig. 14 shows the resonant region of final photon frequency values ω_i (in units of the initial electron energy E_i), which is determined by the system of the equations and the inequalities (215), (228), (230), as a function of the parameter $\alpha = (\theta - \tilde{\theta})\,(m/E)$ when $\omega = 2.36$ eV, $E_i = 48.0$ GeV (since $E_i \gg \omega$, then $\tilde{\theta}_S \approx \tilde{\theta}$), $\theta = 163°$.

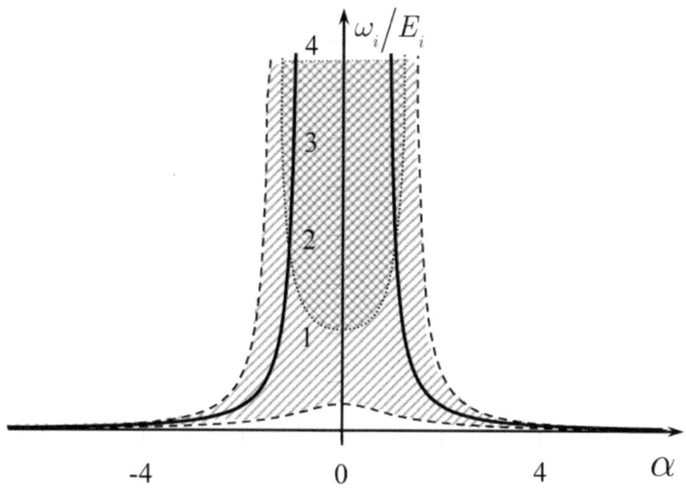

resonance of direct diagram (line thickness ~ φ_0^{-1});
resonance of exchange diagram through electron state;
resonance of exchange diagram through positron state.

Fig. 14. Resonant region of frequencies $\omega_i(\alpha)$ of an ingoing photon (in units of the ingoing electron energy E_i), which is determined by the system of equations and inequalities (215), (228), (230).

5.3 Resonant probability for the direct diagram

We consider the case when the conditions of the direct diagram resonance (220) are realized. Thought it is accompanied by the exchange diagram resonance, but its contribution may be neglected in the following cases:

1. when an initial photon is emitted out of the strictly defined and narrow region of an initial photon directions when the exchange diagram resonance occurs (see Fig. 14);

2. when the total probability is obtained, since the contribution to the total probability from the exchange diagram is $\sim (\omega\tau)^{-1} \ll 1$ and, therefore, it may be neglected.

The differential probability is obtained by standard mode (Berestetskii et al. (1982)). After averaging over initial particle polarizations and summation over final particle polarizations and also the integration over frequencies ω_f and the azimuthal angle $\psi' = \angle\left(\mathbf{e}_x, \mathbf{k}_{f,\perp}\right)$ of a final photon emission we obtain the differential probability:

$$dW_{fi}^{res} \approx \frac{2e^4\eta_0^4 m^2}{\pi\omega_i E_i V \tilde{u}_1}\,(\omega\tau)^2\,P_{\text{res}}(\beta)\left[f(u',\tilde{u}_1)f(u,\tilde{u}_1) - g(u',\tilde{u}_1)g(u,\tilde{u}_1)\right]\frac{du'}{(1+u')^2}\tau. \quad (232)$$

Here,

$$
\begin{aligned}
u &= \frac{(kk_i)}{(qk_i)},\\
u' &= \frac{(kk_f)}{(qk_f)},\\
\tilde{u}_1 &= \frac{u_1}{1-\tilde{u}},\\
u_{\mathrm{res}} &= \frac{\tilde{u}}{1-\tilde{u}},
\end{aligned}
\tag{233}
$$

at that $0 \le u \le \tilde{u}_1$, $0 \le u' \le \tilde{u}_1$, $0 \le \tilde{u}' \le \tilde{u}_1$. In Eq. (232) $P_{\mathrm{res}}(\beta)$ is the function, which determines the resonant profile (see Fig. 15). It is obtained by

$$
P_{\mathrm{res}}(\beta) = \frac{1}{2\pi}\int_{-\infty}^{\infty} |I_1(\beta, l_*)|^2 d(\varphi_0 l_*). \tag{234}
$$

We determine the resonance width at a half of the probability maximum (see Fig. 15). The

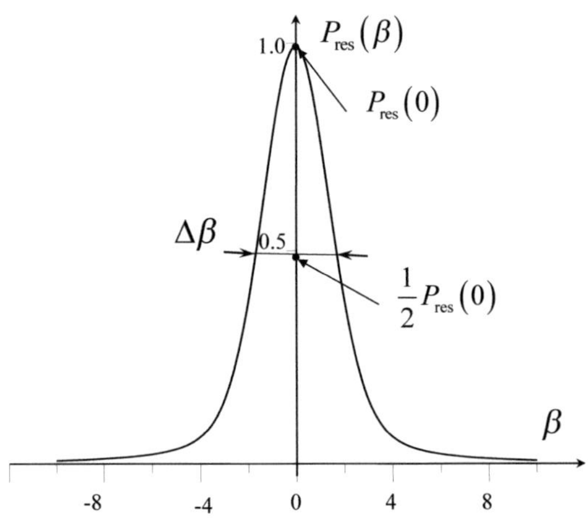

Fig. 15. Dependence of the function P_{res} (234), which determines the resonant profile, on the resonant parameter β (221).

width which corresponds to the resonant parameter β is equal to $\Delta\beta \approx 3.40$. Therefore, the width specified by the field pulsed character is obtained by

$$
\Gamma_{\mathrm{imp}} = \frac{\Delta\left(q^2 - m^2\right)}{4m} = \Delta\beta \frac{\tilde{u}_1}{2}\frac{m}{\varphi_0} \approx 1.70\frac{m\tilde{u}_1}{\varphi_0}. \tag{235}
$$

We compare the resonance width specified by the field pulsed character (235) with the radiation width:

$$
\Gamma_R = \frac{q_0}{m} W_1 = \frac{e^2 m}{4\sqrt{\pi}}\eta_0^2 F(\tilde{u}_1), \tag{236}
$$

where W_1 is the total probability of the intermediate state decay in a weakly intensive field; the function $F(\tilde{u}_1)$ is defined by

$$F(\tilde{u}_1) = \left(1 - \frac{4}{\tilde{u}_1} - \frac{8}{\tilde{u}_1^2}\right) \ln(1+\tilde{u}_1) + \frac{1}{2} + \frac{8}{\tilde{u}_1} - \frac{1}{2(1+\tilde{u}_1)^2}. \tag{237}$$

This ratio equals

$$\frac{\Gamma_{\text{imp}}}{\Gamma_{\text{R}}} \approx \frac{8.51}{e^2\eta_0^2(\omega\tau)} \frac{\tilde{u}_1}{F(\tilde{u}_1)}. \tag{238}$$

When the condition (199) is met the appraisal value of ratio is equal to $\Gamma_{\text{imp}}/\Gamma_{\text{R}} \geq 10^3 \gg 1$. Therefore, the width specified by the field pulsed character is the major one and the radiation widening may be neglected.

After the integration over the invariant parameter u' we derive the total probability of photon-electron scattering under the direct diagram resonance

$$W_{fi}^{res} \approx \frac{2e^4\eta_0^4 m^2}{\pi\omega_i E_i V \tilde{u}_1} (\omega\tau)^2 P_{\text{res}}(\beta) \left[F(\tilde{u}_1) f(u,\tilde{u}_1) - G(\tilde{u}_1) g(u,\tilde{u}_1)\right] \tau, \tag{239}$$

$$\begin{aligned} G(u',\tilde{u}_1) &= \int_0^{\tilde{u}_1} g(u',\tilde{u}_1) \frac{du'}{(1+u')^2} = \\ &= \frac{1}{4\tilde{u}_1(1+\tilde{u}_1)^2} \left(-4\tilde{u}_1 - 8\tilde{u}_1^2 - 5\tilde{u}_1^3 + (4 + 10\tilde{u}_1 + 8\tilde{u}_1^2 + 2\tilde{u}_1^3) \ln(1+\tilde{u}_1)\right). \end{aligned} \tag{240}$$

Ratio of the total probability (239) to the total probability of the Compton effect in external field absence is expressed as

$$\frac{W_{fi}^{res}}{w_{\text{Compt}}} \approx \frac{\tau}{T} P_{\text{res}}(\beta) \cdot R(u,\tilde{u}_1), \quad R(\tilde{u},u_1) = \frac{2\eta_0^4(\omega\tau)^2}{\pi^2} \frac{\left[F(\tilde{u}_1) f(u,\tilde{u}_1) - G(\tilde{u}_1) g(u,\tilde{u}_1)\right]}{\tilde{u}_1 F(\tilde{u}_1)}, \tag{241}$$

where T is the observation time ($T \gtrsim \tau$), which is determined by conditions of the concrete experiment.

When $u_1 \ll 1$ we derive

$$R(\tilde{u},u_1) \approx \frac{4\eta_0^4(\omega\tau)^2}{\pi^2} \frac{(1-\tilde{u})}{u_1} \left(1 - 2\frac{\tilde{u}}{u_1}\left(1 - \frac{\tilde{u}}{u_1}\right)\right). \tag{242}$$

Fig. 16 demonstrates the ratio of the resonant probability of scattering of a photon by an electron in the field of a pulsed wave to probability of the Compton effect as a function of parameters $\tilde{u}$, u_1 within the resonant peak ($\beta = 0$) under $\tau/T = 1$, $\eta_0 = 0.05$. It can be seen from Fig. 16 that the resonant probability may exceed considerably the probability of the Compton effect in external field absence. This fact becomes apparent particularly in the case $u_1 \ll 1$ (but it should be noticed here, that in view of infrared divergence the formulae (239), (241) are correct within the region $u_1 \gtrsim \eta_0^2$). Within the region $u_1 \gtrsim 1$ this effect disappears. Under conditions $\eta_0^2 \sim u_1 \ll 1$ within the range of optical frequencies $E_i/m \ll m/\omega \sim 10^5$ for the ratio of probabilities is correct $R \sim 10^3$.

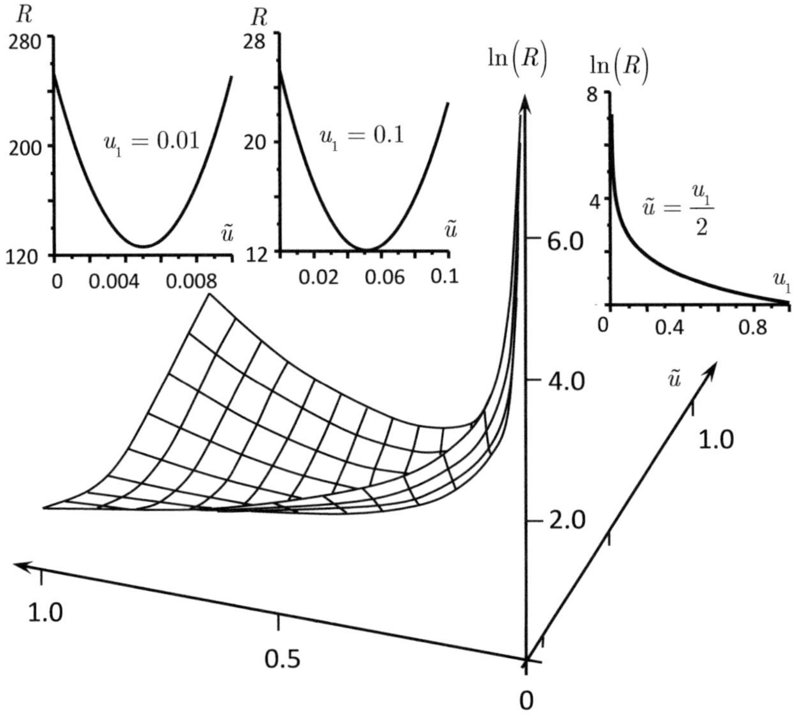

Fig. 16. Ratio of the resonant probability of scattering of a photon by an electron in the field of a pulsed wave to the probability of the Compton effect in external field absence (241) as a function of the parameter u_1 (213) in the resonance peak ($\beta = 0$) under $\tau_{imp}/T = 1, \eta = 0.05$.

6. Conclusions

Performed studies of resonant QED processes in a pulsed light field result:

1. The QED processes of the second order in a pulsed light field may occur under resonant conditions when the four-momentum of an intermediate particle lies near the mass surface.
2. The resonant behavior of the cross-section is specified by characteristics of the laser pulse. The resonant infinity in the process amplitude is eliminated by accounting for the pulsed character of an external field.
3. The differential cross section of the resonant process may be several orders of magnitude higher than the corresponding cross section in external field absence.

The results can be tested in the experiments on verification of quantum electrodynamics in presence of strong fields (SLAC and FAIR).

7. References

Bagnoud, V., Aurand, B., Blazevic, A., Borneis, S., Bruske, C., Ecker, B., Eisenbarth, U., Fils, J., Frank, A., Gaul, E., Goette, S., Haefner, C., Hahn, T., Harres, K., Heuck, H. M., Hochhaus, D., Hoffmann, D., Javorková, D., Kluge, H., Kuehl, T., Kunzer, S., Kreutz, M., Merz-Mantwill, T., Neumayer, P., Onkels, E., Reemts, D., Rosmej,

O., Roth, M., Stoehlker, T., Tauschwitz, A., Zielbauer, B., Zimmer, D. & K.Witte (2009). Commissioning and early experiments of the PHELIX facility, *Appl. Phys B.* 100(1): 137–150.

Baier, V. N., Fadin, V. S. & Khoze, V. A. (1973). Quasi-real electron method in high energy quantum electrodynamics, *Nucl.Phys. B* 65(2): 381–396.

Belousov, I. V. (1977). On resonant Compton scattering in the field of intense electromagnetic wave, *Opt. Com-mun.* 20(2): 205–208.

Berestetskii, V. B., Livshits, E. M. & Pitaevskii, L. P. (1982). *Quantum electrodynamics Volume 4: Course of Theoretical Physics*, Oxford: Pergamon press.

Bhabha, H. J. (1938). On the penetrating component of cosmic radiation, *Proc. Royal Soc. A.* 164: 257–294.

Borisov, A. V., Zhukovskii, V. C. & Eminov, P. A. (1980). Resonant electron-electron bremsstrahlungin the field of an electromagnetic wave, *Sov. Phys. JETP* 51: 267–234.

Borisov, A. V., Zhukovskii, V. C., Nasirov, A. K. & Eminov, P. A. (1981). Resonance twophoton pair production on nuclei and on electrons, *Rus. Phys. J.* 24: 107–110.

Brown, L. S. & Kibble, T. W. B. (1964). Interaction of intense laser beams with electrons, *Phys. Rev.* 133(3A): A705–A719.

Bula, C., McDonald, K., Prebys, E., Bamber, C., Boege, S., Kotseroglou, T., Melissinos, A., Meyerhofer, D., Ragg, W., Burke, D., Field, R., Horton-Smith, G., Odian, A., Spencer, J., D.Walz, andW.M. Bugg, S. B., Shmakov, K. & Weidemann, A. (1996). Observation of nonlinear effects in Compton scattering, *Phys. Rev. Lett.* 76(17): 3116–3119.

Bunkin, F. V. & Fedorov, M. V. (1966). Bremsstrahlung in a strong radiation field, *Sov. Phys. JETP* 22: 844–847.

Burke, D., Field, R., Horton-Smith, G., Spencer, J., Walz, D., Berridge, S., Bugg, W., Shmakov, K., Weidemann, A., Bula, C., McDonald, K., Prebys, E., Bamber, C., Boege, S., Koffas, T., Kotseroglou, T., Melissinos, A., Meyerhofer, D., Reis, D. & Ragg, W. (1997). Positron production in multiphoton light by light scattering, *Phys. Rev. Letts.* 79(9): 1626–1629.

Denisov, M. M. & Fedorov, M. V. (1967). Bremsstrahlung effect on relativistic electrons in a strong radiation field, *Zh. Eksp. Teor. Fiz.* 53: 1340–1348.

Ehlotzky, F., Jaron, A. & Kaminski, J. (1998). Electron-atom collisions in a laser field, *Phys. Rep.* 297(2-3): 63–153.

Ehlotzky, F., Krajewska, K. & Kaminski, J. Z. (2009). Fundamental processes of quantum electrodynamics in laser fields of relativistic power, *Rep. Prog. Phys.* 72(4): 046401.

Gorbunkov, M. V., Konyashkin, A. V. & Kostrukov, G. (2005). Picosecond completely solid-state Nd:YAG-lasers with pulse diode pumping, *Quantum electron.* 35(1): 2–6.

Karapetian, R. V. & Fedorov, M. V. (1978). Spontaneous bremsstrahlung of an electron in the field of an intense electromagnetic wave, *Zh. Eksp. Teor. Fiz.* 75: 816–826.

Klein, O. & Nishina, Y. (1929). Über die Streuung von Strahlung durch freie Elektronen nach der neuen relativistischen Quantendynamik von Dirac, *Z. F. Phys.* 52(11-12): 853–869.

Lebed', A. A. & Roshchupkin, S. P. (2008). The influence of a pulsed light field on the electron scattering by a nucleus, *Laser Physics Letters* 5(6): 437–445.

Lebed', A. A. & Roshchupkin, S. P. (2009). Nonresonant spontaneous bremsstrahlung by a relativistic electron scattered by a nucleus in the field of pulsed light wave, *The Eur. Phys. J. D.* 53(1): 113–122.

Lebed', A. A. & Roshchupkin, S. P. (2010). Resonant spontaneous bremsstrahlung by an electron scattered by a nucleus in the field of a pulsed light wave, *Phys. Rev. A.* 81: 033413(1–13).

Lebed', A. A. & Roshchupkin, S. P. (2011). Resonant electron positron pair photoproduction on a nucleus in a pulsed light field, *JETP* 113(1): 46–54.

Lebedev, I. V. (1972). Generation of harmonics due to electron retardation in the presence of an intensive light wave, *Opt. Spectrosk.* 32: 120–122.

Lötstedt, E., Jentschura, U. D. & Keitel, C. H. (2007). Evaluation of laser assisted bremsstrahlung with Dirac Volkov propagators, *Phys. Rev. Letts.* 98(4): 043002(4 pages).

Lötstedt, E., Jentschura, U. D. & Keitel, C. H. (2008). Laser channeling of bethe-heitler pairs, *Phys. Rev. Lett.* 101: 203001 (1–4).

Massey, H. S. W. & Corben, H. C. (1939). Elastic collisions of meson with electrons and protons, *Proc. Camb. Phil. Soc.* 35(3): 463–473.

Möller, C. (1932). Zur Theorie des Durchgangs schneller Elektronen durch Materie, *Ann. Phys.* 406(5): 531–585.

Narozhniy, N. & Fofanov, M. S. (1996). Photon emission by an electron in a collision with a short focused laser pulse, *Journal of Experimental and Theoretical Physics* 83: 14–24.

Narozhny, N. B. & Fofanov, M. S. (1997). Creation of a pair by a photon colliding with a short focused laser pulse, *Laser Phys.* 7(1): 141–149.

Nikishov, A. & Ritus, V. (1979). Quantum electrodynamics of phenomena in an intense field, *Trudy FIAN* 111: 1–279.

Oleinik, V. (1967). Resonant effects in the field of intense laser beam, *Journal of Experimental and Theoretical Physics* 52: 1049–1067.

Padusenko, E. A. & Roshchupkin, S. P. (2010). Resonant scattering of a lepton by a lepton in the pulsed light field, *Las. Phys.* 20(12): 2080–2091.

Padusenko, E. A., Roshchupkin, S. P. & Voroshilo, A. I. (2009). Nonresonant scattering of relativistic electron by relativistic muon in the pulsed light field, *Laser Physics Letters.* 6(3): 242–251.

Roshchupkin, S. P. (1983). Resonance effects with the photoproduction of electron-positron pairs in the field of a plane electromagnetic wave, *Rus. Phys. J.* 26(8): 683–686.

Roshchupkin, S. P. (1985). Bremsstrahlung of a relativistic electron scattered by a nucleus in a strong electromagnetic field, *Yad. Fiz.* 41: 1244–1257.

Roshchupkin, S. P. (1996). Resonant effect in collisions of relativistic electrons in the field of light wave, *Laser Phys.* 6(5): 837–858.

Roshchupkin, S. P., Tsybul'nik, V. A. & Chmirev, A. N. (2000). Probability of multiphoton processes in phenomena of a quantum electrodynamics in a strong light field, *Laser Physics.* 10(6): 1231–1248.

Roshchupkin, S. P. & Voroshilo, A. I. (2008). *Resonant and Coherent Effects of Quantum Electrodynamics in the Light Field [in Russian]*, Naukova Dumka, Kiev.

Schnez, S., Lötstedt, E., Jentschura, U. D. & Keitel, C. H. (2007). Laser assisted bremsstrahlung for circular and linear polarization, *Phys. Rev. A.* 75(5): 53412 (12 pages).

Schwinger, J. (1951). On gauge invariance and vacuum polarization, *Phys. Rev.* 82(5): 664–679.

Volkov, D. (1935). Über eine Klasse von Lösungen der Diracschen Gleichung, *Z. Phys.* 94: 250.

Voroshilo, A. I. & Roshchupkin, S. P. (2005). Resonant scattering of a photon by an electron in the field of a circularly polarized electromagnetic wave, *Laser Phys. Letts.* 2(4): 184–189.

Voroshilo, A. I., Roshchupkin, S. P. & Nedoreshta, V. N. (2011). Resonant scattering of photon by electron in the presence of the pulsed laser field, *Laser Physics.* 21(9): 1675–1687.

Zhou, F. & Rosenberg, L. (1993). Bremsstrahlung in laser-assisted scattering, *Phys. Rev. A* 48: 505–515.

6

Cold Atoms Experiments: Influence of Laser Intensity Imbalance on Cloud Formation

Ignacio E. Olivares and Felipe A. Aguilar
Universidad de Santiago de Chile
Chile

1. Introduction

Following a number of initial experiments in a magneto optical trap published by us in the period 2008-2009 (Olivares et al, 2008, 2009), there has been an increase in activity in the field. A brief review of the experimental methods can be found in (Olivares, 2007, 2008; Milonni, 2010). The physical details needed to obtain a cloud of cold atoms were described by Metcalf (1999). We will survey the literature and make a thorough discussion of the conditions that permit a stable cloud. We will outline our approach to the construction of the magneto optical trap. Our experiments are based on the construction described by Wieman et al. (1998) and Rapol et al. (2001). We followed the guidelines given in these articles but used state of the art equipment to obtain reliable results in our initial attempts to obtain a cloud of cold atoms. The only initial exception was a self made optical glass cell that was considered inexpensive. Subsequently, it was replaced by a more technically advanced cell that permitted us to improve the observational capability of the system. We will describe an experiment that proved the stability of the cloud and the optical method to vary the laser intensity of the pump and trap beams. We will study the influence of laser intensity imbalance on cloud formation and give values for the threshold intensity of each laser that supports cloud formation.

2. Description of saturated absorption spectroscopy

Saturated absorption spectroscopy is a simple technique to measure the narrow-line atomic spectral feature limited only by the natural linewidth, that is typically 6 MHz or less (Milonni, 2010). A strong laser beam called the pump beam is directed through an optical cell that contains a vapour as shown in Fig.1. A small part of the pump beam used as a probe beam is sent through the cell in the counter-propagating direction and detected by a simple photodiode.

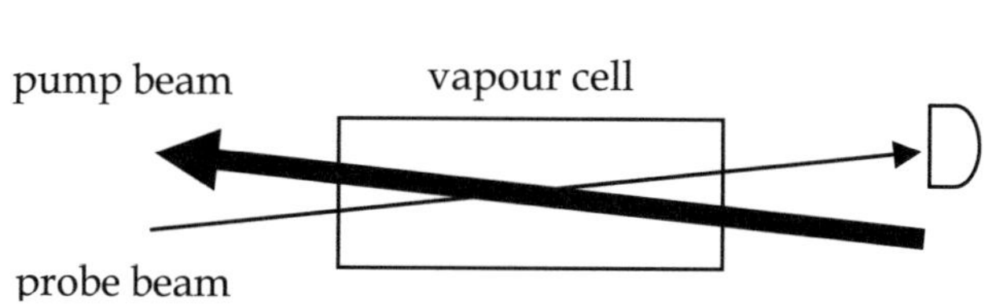

Fig. 1. Basic setup for saturation absorption spectroscopy.

The probe beam can be disposed at a small angle or collinear with respect to the pump beam. The laser frequency is scanned close to the atomic resonance. In the case of a two-level atom system the spectral feature looks like Fig.2. The upper feature is the detected absorption feature when the pump beam is blocked. It shows a Doppler-broadened line which is much broader than the natural linewidth. In the case of weak absorption the feature is a Gaussian profile. The atoms in the vapour move with different velocities in different directions following the Boltzmann velocity distribution. Considering the velocity component of the atoms along the probe beam we have that some atoms move with velocity component in the same direction as the probe beam propagation and other in the opposite direction. The lower feature is the detected intensity with pump laser (Fig.3). It shows a spike just at the atomic resonance frequency ν_0. This spike is known as Lamb dip. When the laser is tuned at $\nu_0 - \Delta\nu$, it will be absorbed only by atoms moving toward the probe laser with longitudinal velocity $\upsilon = c\Delta\nu / \nu_0$. The beam will not be absorbed by atoms with different longitudinal velocities because they are not in resonance so they don't contribute to absorption. Atoms with zero velocity absorb light from the pump laser and become saturated. The probe laser moves through a saturated transparent group of atoms reducing the absorption and producing the Lamb dip.

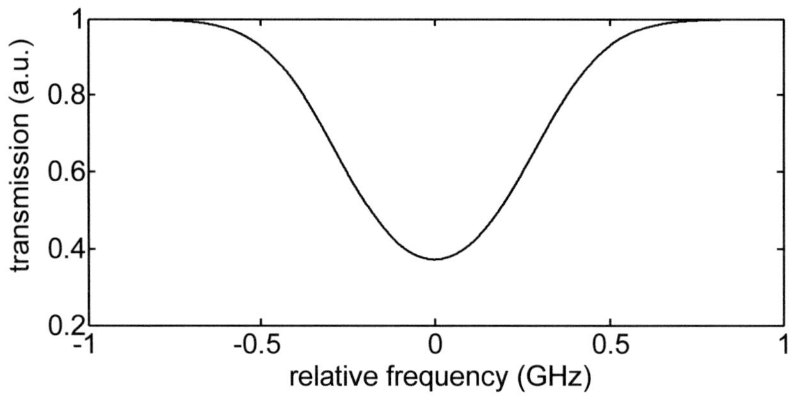

Fig. 2. Absorption line.

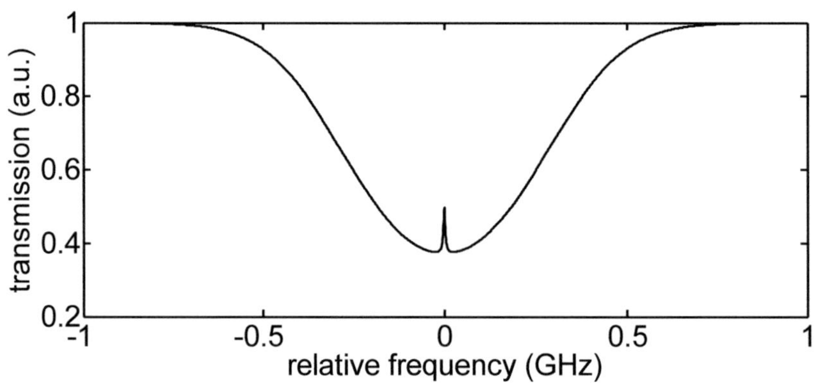

Fig. 3. Doppler free saturated absorption line.

2.1 Multilevel atoms

In the case of a three level system with two closely spaced upper levels and one ground level the spectral features presents two ordinary Lamb dips at the resonance frequencies of the associated transitions and one cross over peak situated just between these two dips at the average of these frequencies as shown in Fig.4. When the laser is tuned at the cross over frequency it is absorbed by one transition from atoms moving toward the laser and by the other transition of the same atom by the laser beam oriented in the opposite direction. The increase of the population of the upper level caused by the strongest laser (pump beam) produces an increase of the transmission of the probe beam at the cross over frequency.

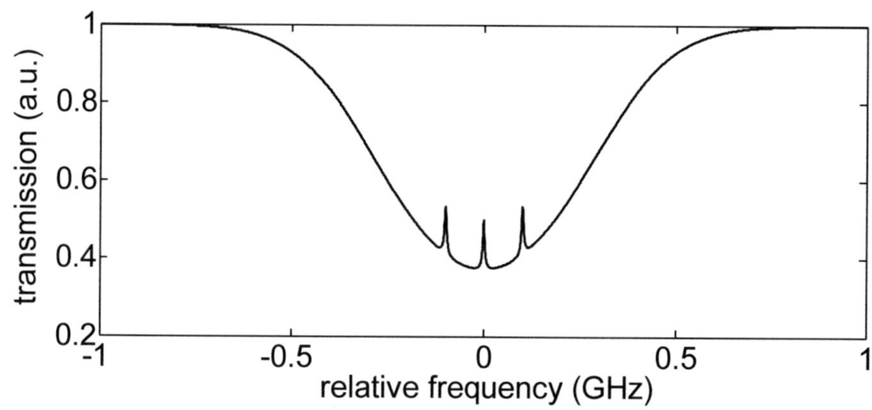

Fig. 4. Positive cross over.

When the system has two closely spaced ground levels and one single upper level the cross over is still half between the ordinary Lamb dips but it exhibits a reduction of transmission due a process named "optical pumping" (Fig.5). Here the laser is absorbed by one optical transition from atoms moving toward the laser. The atoms decay to the second ground level producing an increase of absorption of the probe laser beam driven in the opposite direction.

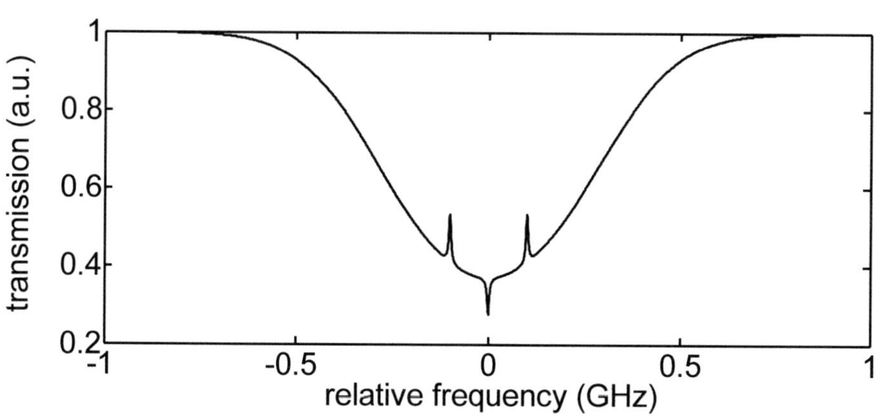

Fig. 5. Negative cross over.

2.2 The saturated absorption spectrometer

The optical setup of the saturated absorption spectrometer is depicted in Fig.6. The signal obtained by the photodiode PD1 can be used as a reference for the Doppler limited spectra.

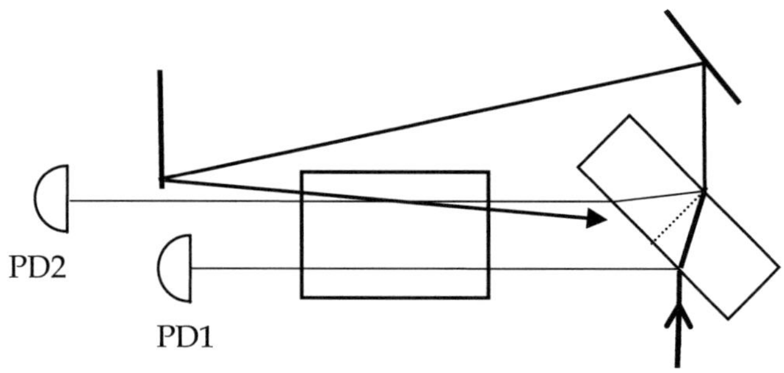

Fig. 6. Saturated absorption spectrometer. The pump beam is indicated with a broader line. The signal obtained by the photodiode PD2 contains the Doppler free feature.

2.3 Semiquantitative ideas at two level atoms

The saturated absorption spectra can be calculated with a simplified model for two level atoms. The differential contribution to the absorption coefficient by atoms with velocity between υ and $\upsilon + d\upsilon$ can be written as

$$d\tau(\nu,\upsilon) = \tau_0 \lambda_0 (P_1 - P_2) F dn \qquad (1)$$

where τ_0 is the optical depth at the centre of the resonance, P_1 and P_2 are the relative populations of the ground and excited state respectively,

$$F(\upsilon,\nu) = \frac{\Gamma / 2\pi}{(\nu - \nu_0 + \nu_0 \upsilon / c) + \Gamma^2 / 4} \qquad (2)$$

is the normalized Lorentzian absorption profile for atoms with natural linewidth Γ including the Doppler shift and

$$dn \propto e^{-m\upsilon^2/kT} d\upsilon \qquad (3)$$

the Boltzman distribution for velocities along the beam axis. The transmission of the probe beam through the cell is $e^{-\tau(\nu)}$. In the case that the pump laser is turned off and the probe laser beam intensity is low we have that few atoms will be excited and most of the atoms will remain in its ground state. In this case $P_2 = 0$ and $P_1 = 1$. For example in the case of rubidium when $\tau_0 = 1$, T = 300ºK, and $\Gamma = 6$ MHz we obtained by numerical integration of Ec. 2 the profile shown in Fig.1. To obtain the relative populations of the ground and excited states when the system is illuminated by the strong pump laser it is neccesary to write the rate equation for a two level system as

$$\dot{P}_1 = \Gamma P_2 - \frac{1}{c} I_p S(\nu)\left(B_{12}P_1 - B_{21}P_2\right) \tag{4}$$

where Γ corresponds to the excited lifetime, I_p is the intensity of the pump laser and

$$B_{21} = \frac{c^3}{8\pi h \nu_0^3}\Gamma \tag{5}$$

is the stimulated emission coefficient,

$$B_{12} = (g_1 / g_2)B_{12} \tag{6}$$

the absorption coeficient, g_1 and g_2 are the degeneracy's of the ground and excited states respectively, and

$$S(\nu) = \frac{\Gamma / 2\pi}{\delta^2 + \Gamma^2 / 4} \tag{7}$$

the atom lineshape, with $\delta = (\nu - \nu_0 - \nu_0 \upsilon / c)$. The minus sign is explained because the pump laser is in the counterpropagating direction in relation to the probe laser. In stationary state we have $\dot{P}_1 = \dot{P}_2 = 0$ and as $P_1 + P_2 = 1$ we have

$$P_1 - P_2 = 1 - 2P_2 \tag{8}$$

Solving Ec. 4 for P_2 in stationary state and assuming that $g_1 = g_2$ we have that

$$P_2 = \frac{s/2}{1 + s + 4\delta^2 / \Gamma^2} \tag{9}$$

where $s = I / I_{sat}$, I the intensity of the pump laser and $I_{sat} = 2\pi^2 hc\Gamma / \lambda^3$ is the saturation intensity. To plot a profile with one single Lamb dip we used the calculated excited population from Ec.9. For example, Fig. 2 was obtained integrating numerically the transmission coefficient for rubidium atoms with $\lambda_0 = 780$ nm, $\tau_0 = 1$, $T = 300$ K and $\Gamma = 6$ MHz and considering the pump laser.

2.4 Energy level diagram

The energy level diagram (Fig. 7) contains two ground hyperfine levels separated by nearly 3 GHz and four excited levels separated by less than the Doppler broadened line. As the atoms pumped by the cooling laser from the F = 3 level into the F′ = 4 level decay into the F = 2 level it is necessary to optically pump the atoms from this level back to the F = 3 level through the F′ = 3 level. This is done by the repumping laser.

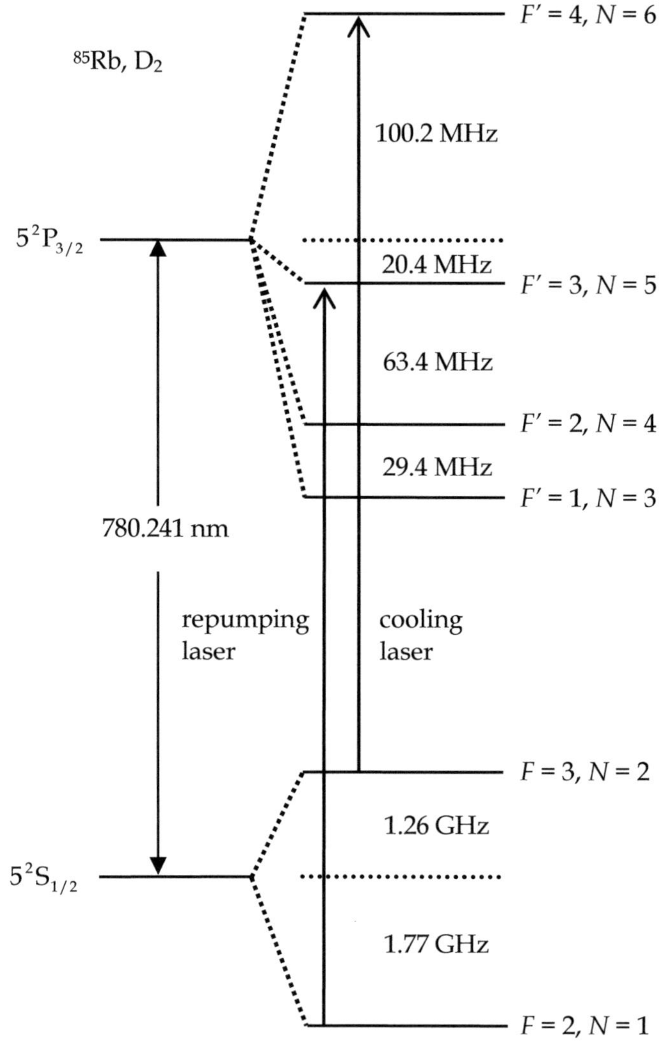

Fig. 7. Energy level diagram. The transitions for cooling and optical repumping are indicated with arrows.

3. Detailed saturated absorption using density matrix elements

The transition rate is given by

$$W_{jk} = \frac{\frac{1}{2}\cdot\left(\frac{8\pi K_e}{c\hbar^2}\|D\|^2\right)\cdot\Gamma}{\left((\omega-\omega_0)-(\omega_j-\omega_k)-kv\right)^2+\Gamma^2}\cdot N_{jk}\cdot I \tag{10}$$

with $N_{13} = 1/9$, $N_{14} = 7/81$, $N_{15} = 4/81$, $N_{24} = 2/81$, $N_{25} = 5/81$ and $N_{26} = 1/9$, I the laser intensity,

$$\|D\|^2 = \frac{1}{\tau} \cdot \frac{3\hbar c^3 (2L_f + 1)}{4\omega^3 K_e} \tag{11}$$

is the square of the reduced matrix element, $K_e = 8.99 \times 10^9$ Vm/C the Coulomb constant, τ the lifetime of the excited atoms and $L_f = 1$. The optical Bloch equations for the relative populations ρ_{11} to ρ_{66} of the ^{85}Rb D_2 line are given by

$$\begin{aligned} \dot{\rho}_{11} &= W_{13}\left(\rho_{33} - \frac{g_3}{g_1} \cdot \rho_{11}\right) + W_{14}\left(\rho_{44} - \frac{g_4}{g_1}\rho_{11}\right) + W_{15}(\rho_{55} - \frac{g_5}{g_1}\rho_{11}) \\ &+ \gamma\left(\rho_{33} + \frac{7}{9}\rho_{44} + \frac{4}{9}\rho_{55}\right) + \gamma_T\left(\frac{5}{12} - \rho_{11}\right) \end{aligned} \tag{12}$$

$$\begin{aligned} \dot{\rho}_{22} &= W_{24}\left(\rho_{44} - \frac{g_4}{g_2} \cdot \rho_{22}\right) + W_{25}\left(\rho_{55} - \frac{g_5}{g_2}\rho_{22}\right) + W_{26}\left(\rho_{66} - \frac{g_6}{g_2}\rho_{22}\right) \\ &+ \gamma\left(\frac{2}{9}\rho_{44} + \frac{5}{9}\rho_{55} + \rho_{66}\right) + \gamma_T\left(\frac{7}{12} - \rho_{22}\right) \end{aligned} \tag{13}$$

$$\dot{\rho}_{33} = W_{13}\left(\frac{g_3}{g_1}\rho_{11} - \rho_{33}\right) - (\gamma + \gamma_T)\rho_{33} \tag{14}$$

$$\dot{\rho}_{44} = W_{14}\left(\frac{g_4}{g_1}\rho_{11} - \rho_{44}\right) + W_{24}\left(\frac{g_4}{g_2}\rho_{22} - \rho_{44}\right) - (\gamma + \gamma_T)\rho_{44} \tag{15}$$

$$\dot{\rho}_{55} = W_{15}\left(\frac{g_5}{g_1}\rho_{11} - \rho_{55}\right) + W_{25}\left(\frac{g_5}{g_2}\rho_{22} - \rho_{55}\right) - (\gamma + \gamma_T)\rho_{55} \tag{16}$$

$$1 = \rho_{11} + \rho_{22} + \rho_{33} + \rho_{44} + \rho_{55} + \rho_{66} \tag{17}$$

where $g_1 = 5$, $g_2 = 7$, $g_3 = 3$, $g_4 = 5$, $g_5 = 7$ and $g_6 = 9$, the levels labeled with N = 1 and 2 corresponds to the ground states and the levels labeled with N = 3 to 6 are the excited states, $\gamma_T = \upsilon / d$ is the transit time broadening with d the diameter of the beam and υ the average velocity of the atoms along the beam diameter, $\gamma' = \gamma + \gamma_T$, $\Gamma = \gamma / 2 + \gamma_T$ and $\rho_e = \rho_{33} + \rho_{44} + \rho_{55} + \rho_{66}$ is the total population of the various excited states. In stationary state the time derivatives of the relative popuations become zero. The absorption of the laser light in a vapor with density n and length dx

$$dI = h\,\nu_{if}\, n \left\langle \sum_{if} W_{if}\left(\rho_{ff} - \rho_{ii}\right) \right\rangle dx = -h\,\nu_{if}\, n\gamma' \langle \rho_e \rangle dx \tag{18}$$

The angle brackets indicate the average over the velocity distribution for vapor at temperature T, given by

$$F(\upsilon)=\left(\frac{m}{2\pi k_B T}\right)^{1/2}\exp\left(-\frac{m\upsilon^2}{2k_B T}\right) \tag{19}$$

Extending the absorption equations to te Doppler-free saturation spectroscopy we have

$$dI = h\,\nu_{if}\,n\left\langle \sum_{if} W_{if}^{+}\left(\rho_{ff}^{-}-\rho_{ii}^{-}\right)\right\rangle dx \tag{20}$$

where the population depends on the transition rate $W^{+}=W\left(I_d,\upsilon\right)$ determined by the probe beam with intensity I_d and the transition probability $W^{-}=W\left(I_p,-\upsilon\right)$ due to the pump beam with intensity I_p propagating in the opposite direction, and $\rho_{ii}^{-}=\rho_{ii}\left(W^{+}+W^{-}\right)$.

4. Experimental details

The experiment was installed in a 6x12 feet optical top [1] that was passively damped. The experiment included two tuneable diode lasers, two saturated absorption spectrometers, two scanning interferometers, a complete vacuum system, beam expanders, polarizing optics, infrared camera, optics and mechanics components, a rubidium cells, and photodiodes.

4.1 The saturated absorption spectrometer

The saturated absorption spectrometer is shown in Fig.8. The laser beam was lifted 15 cm above the optical top level by the mirrors M1 and M2, and directed to the first optical glass beam divider. A small part of the beam was directed to the second optical glass divider, the strongest beam went to the trap. The second beam divider drives the strongest beam to the interferometer and the small beam act as a pump laser in the rubidium cell. The beam reflected off the mirror M3 acts as a test weak beam that was measured by a photodiode [2].

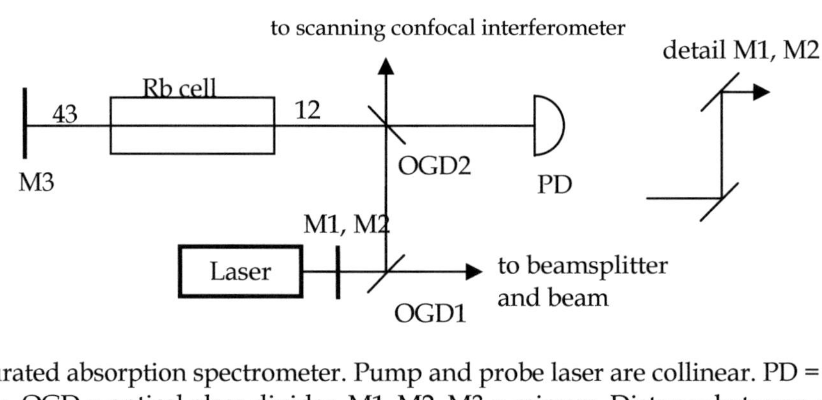

Fig. 8. Saturated absorption spectrometer. Pump and probe laser are collinear. PD = photodiode, OGD = optical glass divider, M1, M2, M3 = mirrors. Distance between closest optical components are given in inches.

[1] Thorlabs, Model PTR12114-PTH503
[2] Thorlabs, Model DET10

4.2 Scanning confocal interferometer

The scanning confocal Fabry-Perot interferometer (Fig.9) is a nice tool to check if the laser is running in single mode operation specially. One of the main features of the Fabry-Perot interferometer is that it can measure with high resolution the spectral content of the laser. A basic Fabry-Perot consists of two identical spherical mirrors with radius R separated by a distance L. The use of two curved mirrors is convenient as they permit a good match to the Gaussian beam coming from the laser.

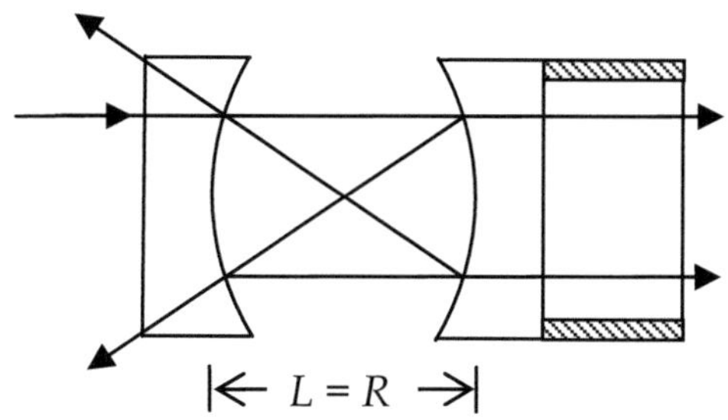

Fig. 9. Confocal scanning Fabry-Perot interferometer.

Two parameters defines the properties of a Fabry-Perot, the free spectral range and the *finesse* or resolution. The free spectral range (FSR) is defined by

$$FSR = \frac{c}{4nL} \tag{21}$$

where n is the index of refraction of the air between the mirrors, c the speed of light, L the distance between mirrors. Near the centre of the mirrors we have that every time the distance between mirrors is changed by a quarter wavelength $(\lambda/4)$ the same part of the spectrum will be reproduced. The mirrors used in our interferometer [3] have a radius of 75 mm and a FSR = 1GHz. The resolution of the interferometer is given by its *finesse*

$$F^* = \frac{FSR}{\Delta\nu} = \frac{\pi\sqrt{R}}{1-R} \tag{22}$$

where $\Delta\nu$ is the full with at half maximum of the interference maxima and R reflectivity of the mirrors. The *finesse* depends on the mirror reflectivity, the losses due to imperfections on the mirror surfaces or dust, and the alignment of the mirrors. In our interferometer the highest *finesse* reported was larger than F^* = 450. A cylindrical piezoelectric transducer (PZT) is attached to one mirror and can move it in small displacements. To displace the mirror a high voltage is applied between the inner and the outer side of the PZT. The interferometer can be used in scanning mode when the laser wavelength is fixed and the piezo transducer is displaced continuously with a ramp function. In this case it is possible to observe the detailed spectra of the laser. Another option is to scan the laser wavelength with a ramp function and the distance between mirrors remains constant. In this case one can observe the laser spectra and change its absolute position in the oscilloscope by applying a

[3] Toptica Photonics, Model FPI100

constant voltage to the PZT. This option is very useful for finding the resonances needed for cooling.

4.3 Vacuum system

For optimal conditions to form an atomic cloud it is necessary to reach an ultra high vacuum level with pressures lower than 10^{-7} Pa (10^{-9} Torr). Our vacuum system (Fig. 10) was built with pipes with nominal 2.75 inch diameter conflate type flanges made of 308 steel. The connections between the pipes and other devices were sealed with cooper gaskets. Our system consisted in a rotary vane pump, followed by a turbomolecular pump [4] and an ion pump [5]. To measure the low vacuum level up to 1.33×10^{-2} Pa (10^{-4} Torr) we used a Convectron [6] gauge. To measure vacuum pressures lower than 1.33×10^{-3} Pa (10^{-5} Torr) we used a Bayard Alpert gauge [7]. Both gauges were connected to a multi-gauge controller [8]. The ultra high vacuum was measured alternatively with the indicator of ionic pump controller. The vacuum process started with the onset of the rotary vane pump to obtain a vacuum close to 1.33×10^{-2} Pa (10^{-4} Torr). After obtaining this vacuum pressure we started the turbomolecular pump, to obtain a vacuum close to 10^{-5} Pa (10^{-7} Torr). To obtain lower vacuum pressures the system was heated in a process called baking to evaporate the water molecules embedded inside the pipes and chamber. For this we rolled around the pipes and flanges along the vacuum line a heater that was made of a nearly 10 m long AWG26 nichrome wire. To electrically isolate the nichrome wire from the pipes we inserted it into a series of 1 m fiber glass spaghettis that were coupled one by one. To do this we slide the outer part at end of one spaghetti into the inner part of the following. The ionic pump was heated with its own heater, when the pump was switched off. The temperature used in the vacuum process was 120 °C. To reach this temperature we increased the temperature 10 °C every 30 minutes with a Variac transformer by increasing the current along the nichrome wire. The complete baking process took at least 5 days. The first day was used to reach the 120 °C baking temperature. This temperature was kept constant during the next 3 days. In the fifth day we initiated the decrease of the temperature at the same rate as at the heating stage, that is a decrease of 10 °C every 30 minutes. This was a precaution to protect the glass and the glue, because all have different temperature expansion coefficients. To obtain a homogeneous temperature along the vacuum line we made a temperature measurement at different places. For this we installed several thermocouples in some points between the heating wires and the pipes. We also covered the heater with aluminium foil. With the baking of the vacuum line we could reduce the pressure by more than one order of magnitude. The ultimate vacuum was less than 100 nPa (1nTorr).

4.4 Observation optical cell: discussion of different methods

Three versions of observation cells were used in our trap. In the first case we bored a 30 mm hole in the centre of a 2.75 inch conflate type blank flange [9]. On the flat side we constructed a

---

[4] Varian, Model TurboVac V50
[5] Varian, Model VacIon Plus 20 StarCell
[6] Granville Phillips, Model 275238
[7] Varian, 580 Nude ion gauge thoria iridium
[8] Varian, Model L8350301
[9] MDC-Vacuum, Model 110008

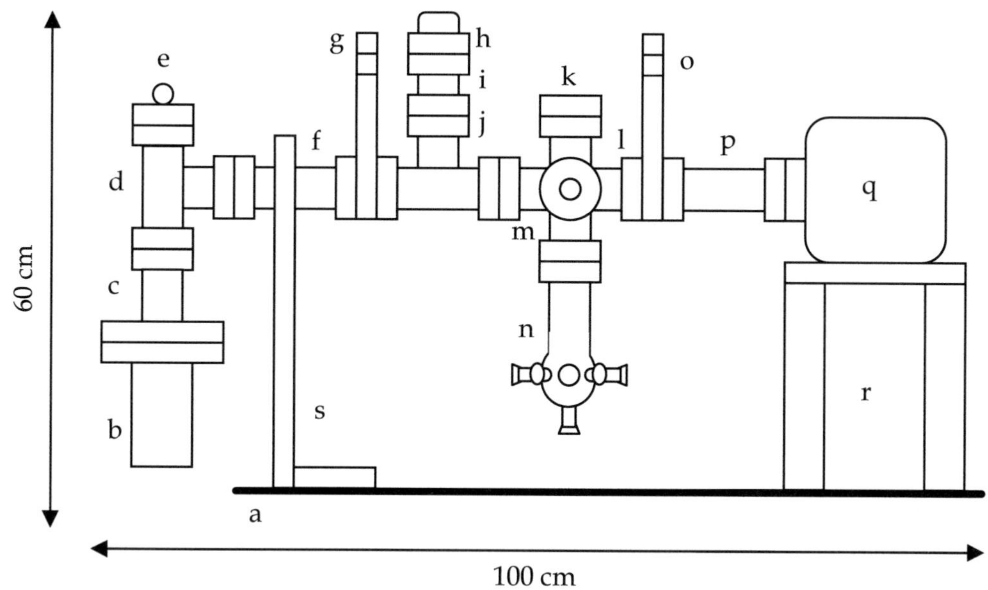

Fig. 10. Vacuum system a: optical table, b: turbomolecular pump, c: reduction nipple CF 4.5 to 2.75 inch [10], d, j: tee CF 2.75 flange [11], e: Convectron vacuum sensor, f, p: nipple [12], g, o: manual valve for ultra high vacuum [13], h: Bayard-Alpert UHV sensor, i: short nipple CF 2.75 flange [14], k: window [15], l: six way cross [16], blank flange for back side [17], m: 8 pins electrical feedthrough, n: bottle with seven horizontal windows and one vertical window, q: ion pump, r: aluminium plate support for ionic pump with dimensions 30x19.5x1 cm mounted in 4 rods of 2 inch diameter, s: L form mount for tubing.

cell that uses four optical glass plates with 4 mm wall thickness and dimensions 35 x 50 mm. On the top of the cell we glued a 35 x 35 mm optical glass plate. The cell was glued to the flat side of the flange. The plates were glued with high vacuum Torr seal [18]. The second version consisted in an optical glass cell with outer wxlxh wall dimensions 55x55x52.5 mm and 2.5 mm wall thickness [19]. The cell was glued on the 4.5 inch side of a zero length reducer from nominal conflate flange 4.5 inch to 2.75 inch. We did not remove the edge of the 4.5 inch side so the cell was installed very tight. This caused that the glass broke after some heat up vacuum procedures. The cell could be repaired several times with the vacuum Torr seal. The first two versions of cells are shown in Fig.11.

[10] MDC-Vacuum, Model 402013
[11] MDC-Vacuum, Model 404002
[12] MDC-Vacuum, Model 402002
[13] MDC-Vacuum, Model 302001
[14] MDC-Vacuum, Model 468008
[15] MDC-Vacuum, Model 450020
[16] MDC-Vacuum, Model 407002
[17] MDC-Vacuum, Model 110008
[18] MDC-Vacuum, Model 9530001
[19] Hellma Cells, Model 704.003-OG

Fig. 11. Left: 55x55x55 mm glass cell, right: glass cell constructed with 35x50x4 mm plates.

The third version (Fig. 12) consisted in a cell prepared by a glass blower. The cell has a 2.75 inch conflate type adapter and 7 optical windows with 1 inch useful area [20].

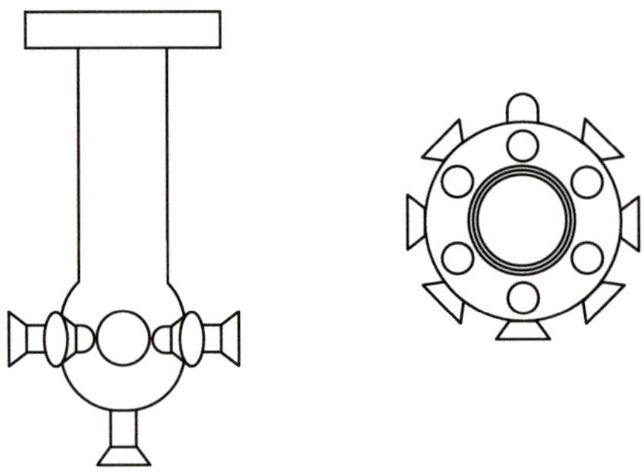

Fig. 12. Side and top view of observation cell.

4.5 Optical layout, detectors and IR camera

The main part of the magneto optical trap optics was purchased as one single item [21]. Our optical layout (Fig.13) include a larger list of parts. The rays coming from each laser are vertically polarized. After leaving the first optical glass divider (OGD1 in Fig.13) each laser beam is driven to a polarizing beamsplitter cube. The polarization of the repumping laser is rotated in 90 degrees by means of a half wave plate and becomes horizontally polarized before entering the polarizing beam splitter cube. The polarization of the cooling laser is maintained vertical and reflected by the beamsplitter cube. By this mean, the cooling laser and the repumping laser become collinear. Both lasers were driven over a line of holes of the optical top and continued collinear at least at 4 meters from the exit of the polarizing beam splitter cube. The polarization of the cooling laser was orthogonal to the polarization of the

[20] MDC-Vacuum, Model 150008

[21] Toptica Photonics, Model MOT-Optics

repumping laser. The combined laser beams were simultaneously expanded by a laser beam expander consisting of a $f = 50$ mm lens followed by two $f = 300$ mm. The diameter of the three lenses was 25 mm. The diameter of the lasers was nearly 3 mm and at the exit it was 12 mm giving an expansion of 4x. The laser disk was rounded by an iris diaphragm.

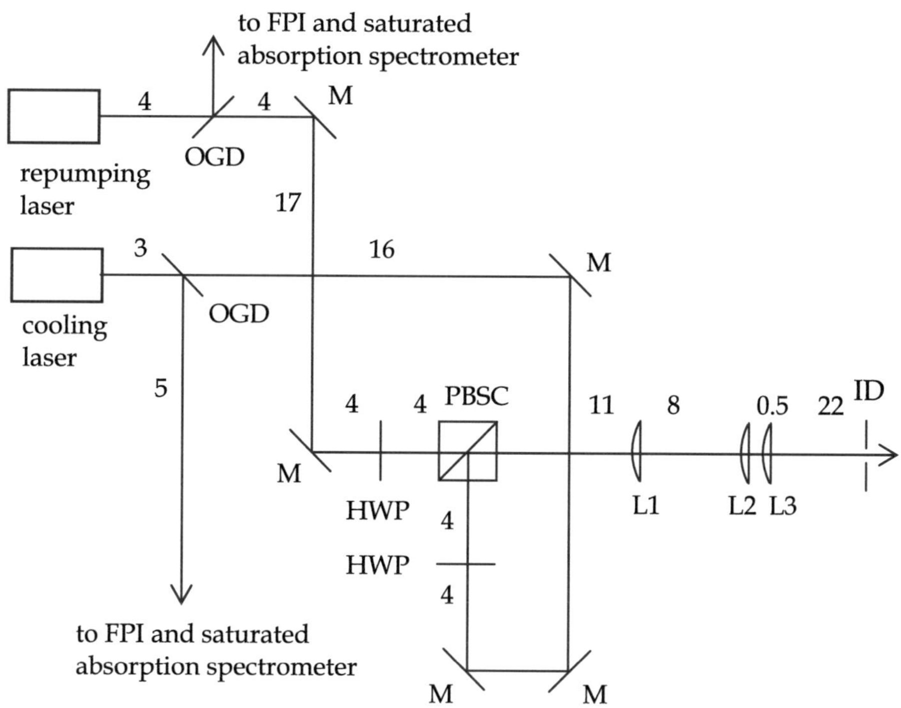

Fig. 13. Combination of repumping and cooling laser beams followed by simultaneous beam expansion. OGD = optical glass divider, L = lenses, HWP = half wave plate, PBSC = polarizing beam splitter cube, ID = iris diaphragm.

After passing the iris diaphragm, both lasers were divided in a 0.3/0.7 divider. Most of the laser power (70%) was directed to the horizontal plane (Fig. 14). A polarizing beam splitter cube divided both lasers equally. Each pair of beams that leaved the polarizing beam splitter cube were divided again by means of two non polarizing beam splitter cubes. By this method it was possible to obtain two sets of counter propagating pairs of beams. In each leg of this arrangements quarter wave plates to with the correct circular polarizations. We installed a surveillance IR camera to observe the cloud and an IR CCD [22] with a 50 mm lens.

[22] Altec Vision, Model PL-B771U

The small part of the optical power (0.3 that was obtained at the 0.70/0.30 beam divider was directed vertically to the optical top as shown in Fig. 15 and directed parallel to the horizontal plane to a half wave plate that rotated both lasers in nearly 45º. A polarizing beam splitter cube disposed after the half wave plate divided the beam in two parts with the same intensity. One part went upwards and the other crossed the polarizing beam splitter cube and was directed by means of two mirrors in the counter propagating downward direction. Two quarter wave plates were used to obtain the correct circular polarization. With our experimental conditions we tried to balance the power from every ray as best as possible.

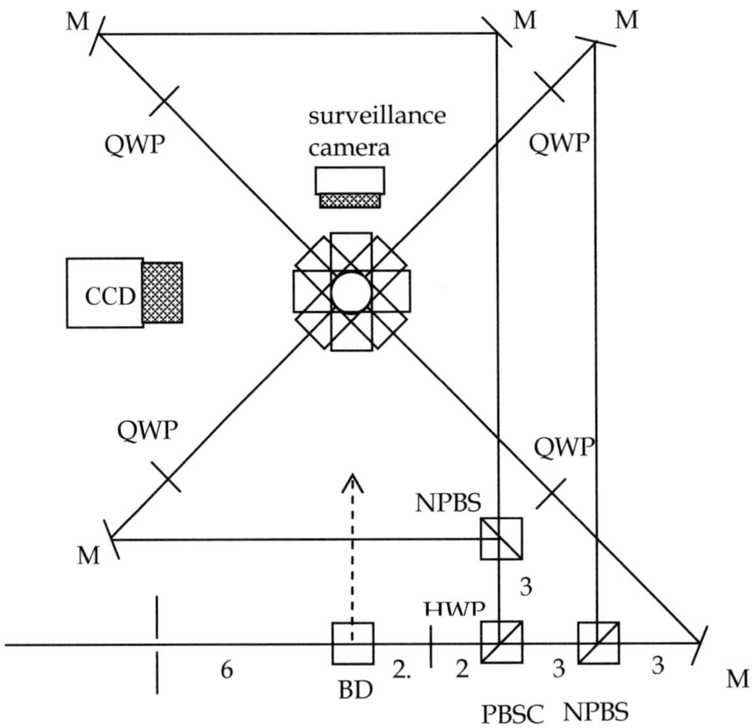

Fig. 14. Beam division in the horizontal plane and use of quarter wave plates to obtain the desired circular polarization. HWP = half wave plate, PBSC = polarizing beam splitter cube, NPBSC = non polarizing beam splitter cube, M = mirror, BD = beam divider 0.3 to vertical plane.

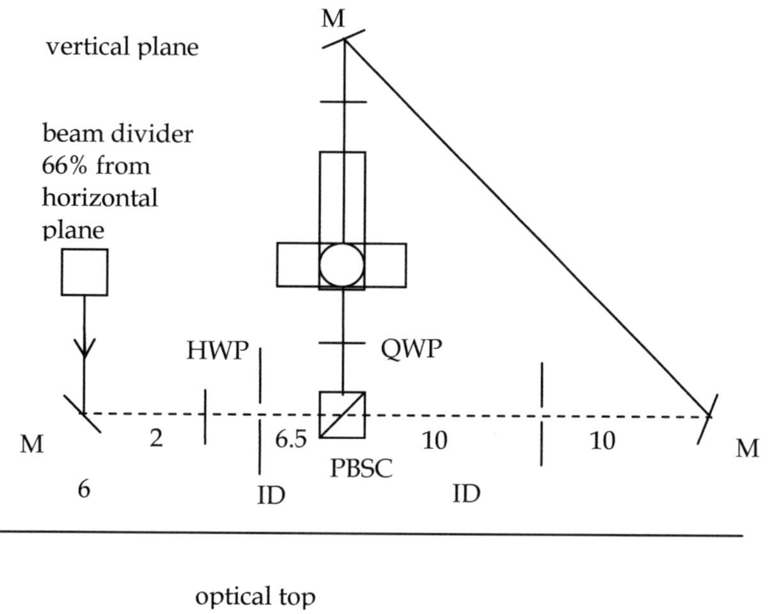

Fig. 15. Beam division in the vertical plane and use of quarter wave plates to obtain the correct circular polarization. QWP = quarter wave plate, PBSC = polarizing beam splitter cube, M = mirror.

4.6 Introduction of neutral atoms using a rubidium getter

A rubidium getter [23] is used to introduce the neutral atoms into the vacuum chamber. The main feature of this getter is that it allows introducing a controlled amount of atoms. The rubidium is released as a vapour when a current flows through the getter. The current required to release the necessary amount of neutral atoms is close to 3.7A. A diagram of the getter is shown in Fig.16. The getter is contained in a chamber with a trapezoidal section and released from a small aperture at the upper part. When the getter cools down, condensation and solidification of the material closes the exit. To start the vapour emission it is necessary to increase the current to 8A during nearly 2 seconds. The pulse duration should be controlled precisely by means of a programmable current power supply [24] to avoid the destruction by melting of the getter.

The code for the power supply was made with Labview6.0. The code set 5 s at 3 A, 2 s at 8 A, 4 s at 6 A and fixed the current at 3.7 A the rest of the time. Several getters were soldered to pair of pins of an 8 pin conflate flanged power feedthrough [25]. Care was taken to label the

[23] Saes Getters, Model RB/NF/3.4/12 FT10+10
[24] Instek, Model PSM-2010
[25] Kurt K. Lesker, Model EFT0084033

earths. The soldering was made by means of thermocouple point soldering device. This uses three 5.1 mF, 350 V electrolytic capacitors in parallel. 70 V is enough to sold the parts. We used only one getter for more than 100 hours and it is still working.

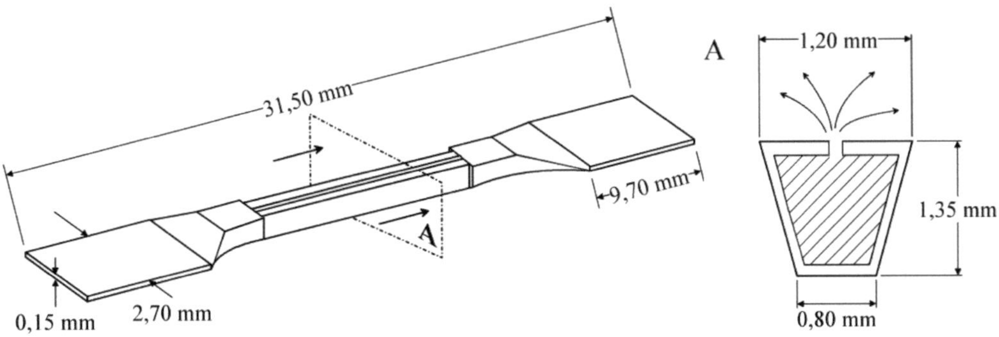

Fig. 16. View of the rubidium getter.

4.7 The pump and the probe laser

The pump and probe lasers used in our experiments are *Littrow* cavity diode lasers [26] delivering about 50 mW of single longitudinal mode emission near 780 nm at a laser line width of nearly 1 MHz. Each laser was protected with a 60 dB optical isolator [27]. The optical isolators were placed inside the laser case after the grating. The use of the optical isolators is essential to obtain a reproducible magneto optical trap as it is very difficult to avoid reflections back into the laser. These reflections can destroy the single mode emission of the lasers.

4.8 Description of the Pound Drever Hall method for frequency stability of the pump and probe lasers

The setup of a cold atom cloud requires fixed cooling and repumping laser frequencies. It is possible to obtain the cloud of cooled atoms without stabilizing the laser but it makes the work more difficult. The Pound Drever Hall method permits the stability of the frequency of the laser frequencies close to the resonances. Fig. 17 shows the optical setup of the Pound Drever Hall detector. A diode laser is collimated by a aspheric lens of short focal distance and its wavelength controlled by a grating that reflects its first diffraction order back into the laser cavity. The wavelength is roughly adjusted by rotating the grating. A piezo electric transducer (PZT) can produce fine angular displacements of the grating and control the frequency of the lasers single mode emission at the MHz level. An optical isolator installed in front of the laser permits to avoid unwanted back reflections into the laser cavity. These reflections could destroy the single mode emission of the laser. Laser exiting the optical isolator is driven to the confocal scanning Fabry Perot interferometer. Two mirrors (2M) lifted the laser to 15 cm from the optical top. The beam was conducted by means of an optical glass divider, a mirror and a pair of mirrors that placed the beam at the level of the interferometers axis. The beam passes a polarizing beam splitter cube, a quarter wave plate

[26] Toptica Photonics, Model DL100
[27] TV-Linos, Model FI-790

and was focused with an f = 200 mm lens to the interferometer. The light reflected from the interferometer becomes horizontally polarized after passing twice the quarter wave plate and was reflected by the polarizing beam splitter cube into a fast photodiode [28].

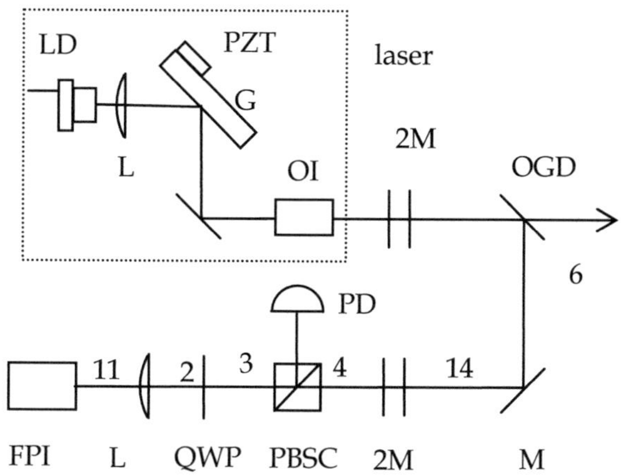

Fig. 17. Optical setup for Pound Drever Hall stabilization method.

The reflected electric field from a Fabry Perot interferometer is given by

$$E_r = \frac{\left(1-e^{i\delta}\right)\sqrt{R}}{1-Re^{i\delta}}E_i \tag{23}$$

where R is the mirror reflectivity and $\delta = 2\pi\Delta\omega / \Delta\omega_{\mathrm{FSR}}$. The laser is modulated at a frequency $\Omega / 2\pi = 20\ \mathrm{MHz}$. The incident laser amplitude can be written as a carrier with two weak sidebands as

$$I(\omega) = J_0^2(\beta)L(\omega;\omega_0) + \sum_{n=1}^{\infty} J_n^2(\beta)\left[L(\omega;\omega_0+n\Omega) + L(\omega;\omega_0-n\Omega)\right] \tag{24}$$

where β is the modulation amplitude,

$$L(w) = \frac{1}{\pi}\frac{\frac{1}{2}\Gamma}{(\omega-\omega_0)^2 + \left(\frac{1}{2}\Gamma\right)^2} \tag{25}$$

is a Lorentzian function, and Γ the laser linewidth. A modulated spectra for $\Omega / 2\pi = 20\ \mathrm{MHz}$ modulation frequency and laser linewidth $\Gamma = 10\ \mathrm{MHz}$ is depicted in Fig.18.

[28] Thorlabs, Model PDA10-EC

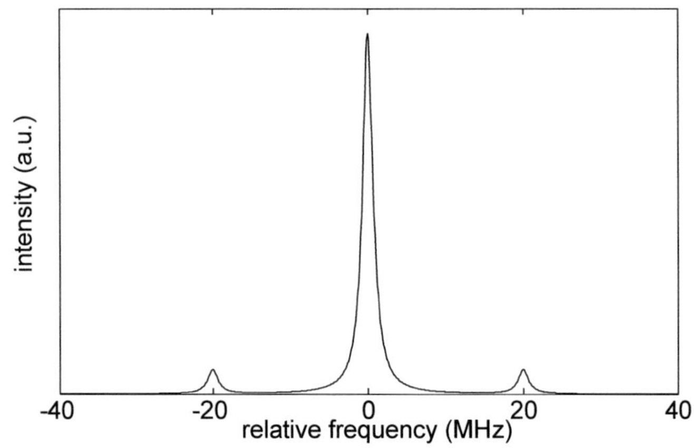

Fig. 18. Laser modulated with 20 MHz sinusoidal function. Sidebands can be seen at both sides of the central feature.

Two sidebands can be found on each side of the central feature. The signal produced by the fast photodiode is mixed with the modulation sinusoidal signal. The error function (Fig.19) is obtained when the product of these two functions is passed through a low pass filter.

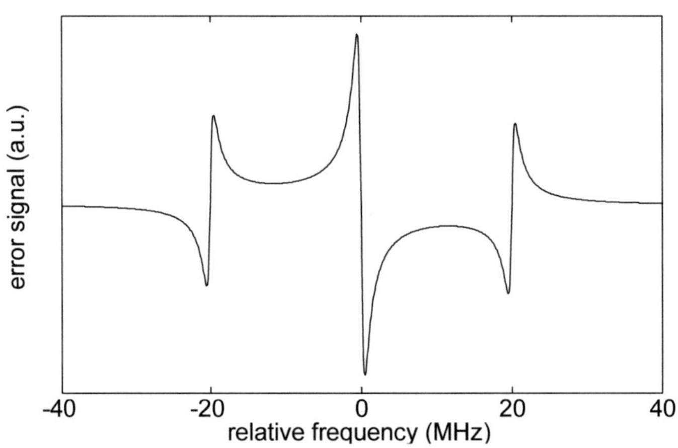

Fig. 19. Error function considering a FSR = 1GHz, finesse = 500 and 20 MHz modulation.

4.9 Polarizing optics: left and right circulating light

Laser beams with opposite helicity polarizations impinge on an atom from opposite directions. Magnetic levels of the atoms are shifted by the magnetic field. The net result is a position-dependent force that pushes the atoms into the center of the magneto optical trap.

In our experiment we used 1 inch diameter multiple order quarter wave plates [29] and 1 inch diameter multiple order half wave plates [30]. The wave plates can be installed in optical

[29] CVI - Melles Griot, Model QWPM-780-10-4

rotating mounts [31] but in our case we used fixed mounts constructed by us. Multiple order wave plates require specification of the used wavelength. In our case the required wavelength was 780 nm. A linear polarized beam incident on a multiple order quarter wave plate produces circular polarized light when the electrical field of the incident laser is oriented at 45 degrees with respect to the optical axis of the quarter wave plate. One of the vector components of the *E*- field is parallel to the optical axis and the other perpendicular. A good method to check the orientation of optical axis of a quarter wave plate is to construct an optical isolator as shown in Fig. 20.

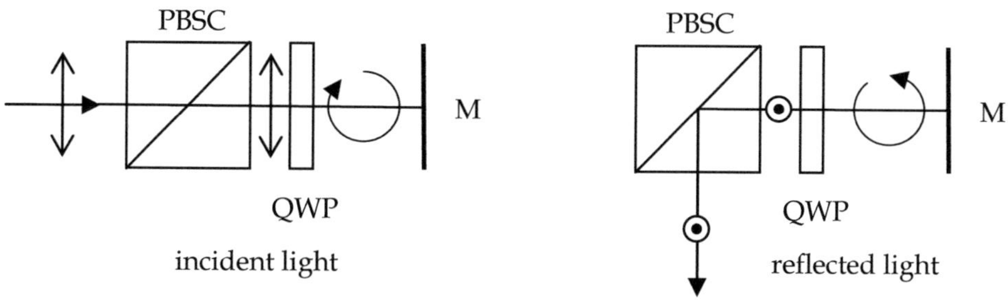

Fig. 20. Optical isolator. PBSC = polarizing beam splitter cube, M = mirror.

Incident light coming from the left side with its polarization vector parallel to the plane of the paper passes the polarizing beam splitter cube and continues through the quarter wave plate. When the optical axis of the plate is rotated in 45° respect to the electrical field the wave becomes circular. The reflected light turns into perpendicular to the plane of the paper and becomes fully reflected by the polarizing beam splitter cube. With the aid of a photodiode it is possible to find the largest reflected signal by rotating the quarter wave plate slightly back and forth. The optical axis of the half wave plate can be found using the linear incident laser light and a polarizing beam splitter cube. When this plate is rotated a 45° relative to the incident field, the field rotates 90°. In general when the half wave plate is rotated at an angle α, the electrical field rotates at an angle 2α. This can be used to obtain a transmission of 0.4 and reflectivity of 0.6. This is correct but the transmitted *E*-fields are slightly rotated to vertical or horizontal.

4.10 Anti Helmholtz coils: magneto optical trap

The force acting on the atoms in the magneto optical trap is position space dependant being larger for atoms that are more distant from the center of the trap. The MOT coils are two copper solenoids with same dimensions and number of windings. The coils are disposed in anti Helmholtz configuration one over the other. Fig. 21 shows a diagram of the coils. The current in one coil flows in opposite direction with respect to the other coil. It is recommended (Wieman, 1995) to have a variable magnetic field gradient with a maximum of 0.2 T/m. We used normally between 0.10 and 0.15 T/m. We used a 1.15 mm diameter (AWG 17), enameled copper wire. Each coil has 196 windings ordered in 14 sheets with 14 windings per sheet. To drive the coils we used two 5A variable current supplies.

[30] CVI - Melles Griot, Model QWPM-780-10-2

[31] Thorlabs, Model RSP1

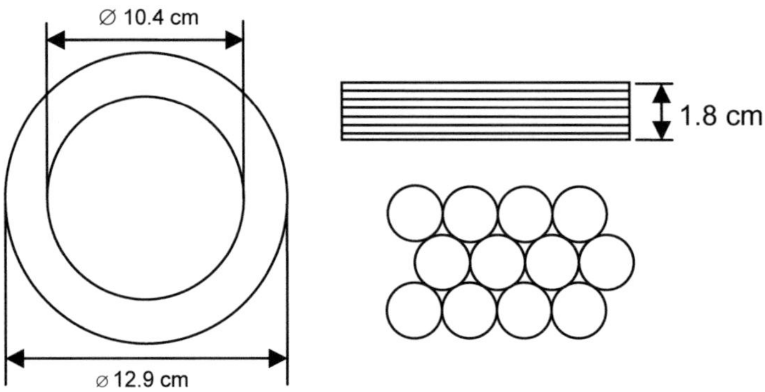

Fig. 21. Construction of anti Helmholtz coils.

5. Finding the spectral lines for repumping and cooling laser

To find the spectral lines for the repumping and cooling laser it is necessary to change the current and temperature of each laser controller and scan the laser piezo element attached at the grating at large amplitudes and measure the whole absorption spectrum from the atoms in the rubidium cell with a photodiode. This should be made for each laser. A typical absorption spectrum of rubidium is shown in Fig.22. Lamb dips are useful to identify the lines.

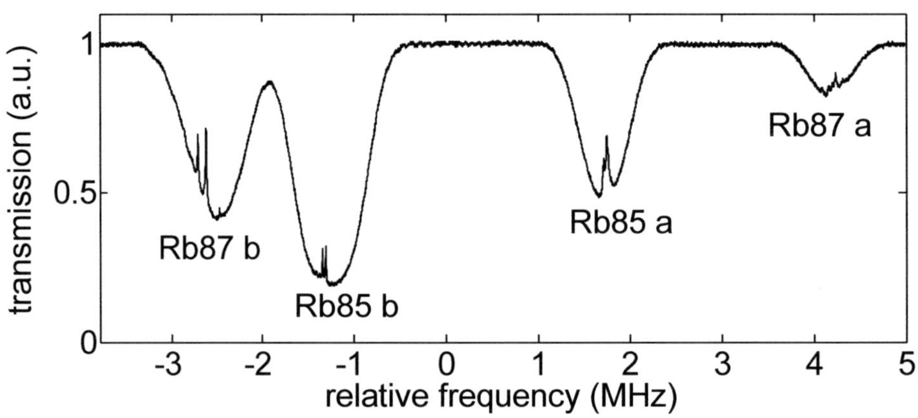

Fig. 22. Saturated absorption spectra used to find spectral lines for the repumping and cooling laser.

6. Doppler free spectra of cooling and repumping laser

A detailed view of the Doppler free spectra for the cooling and repumping lasers is shown in Fig. 23. To obtain these spectra we reduced the scan amplitude of the grating piezo and changed slowly the offset voltage of the piezo to isolate each line. Additionally it was possible to heat the rubidium cell with a nichrome wire to obtain more defined lines.

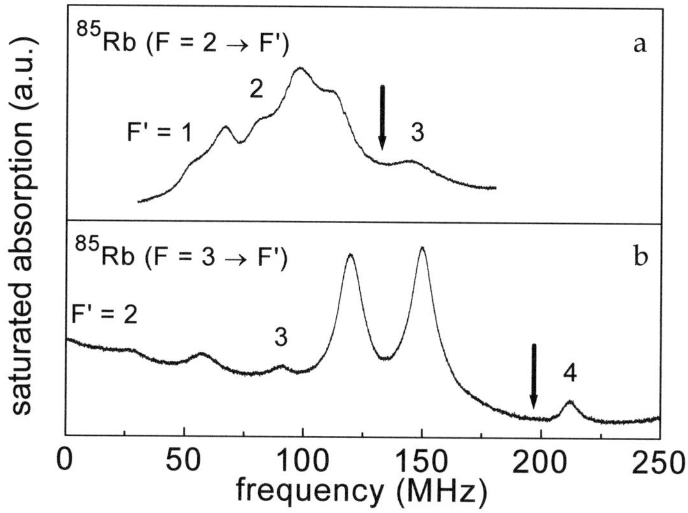

Fig. 23. Doppler free spectra of a) repumping and b) cooling lasers. The arrows indicate the frequencies to be locked.

7. Signals needed to stabilize the repumping and cooling laser

Fig.24 shows a typical measured modulated laser spectra and Fig.25 shows the error function obtained experimentally. In both cases, the interferometer cavity length was held fixed and the laser was scanned continuously. The alignment procedure of the light reflected from the interferometer into the fast photodiode can be best done using a surveillance camera an trying to group the multiple reflections on a single point at the photodiode.

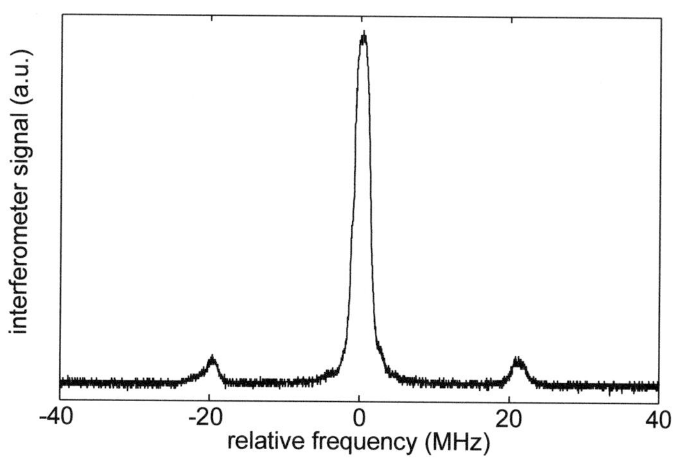

Fig. 24. Laser modulated profile recorded with interferometer.

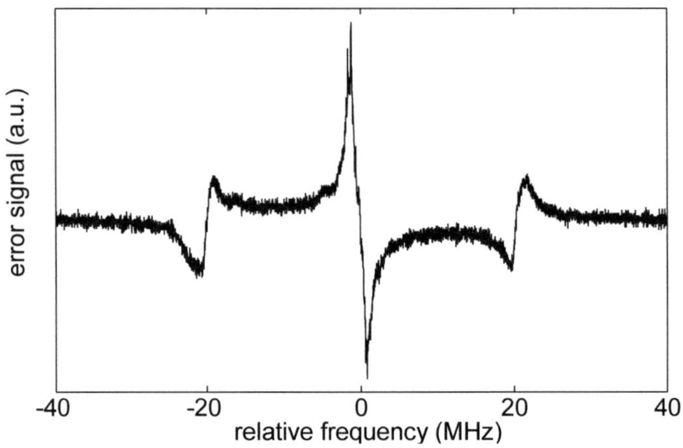

Fig. 25. Experimental error function.

To lock the laser frequency to the needed resonance, we have stored one single Doppler free spectrum and recalled and displayed it on the oscilloscope screen. The amplitude scan was decreased close to the zero crossing of the error function. Adjustments of the error signal position relative to the Doppler free spectra could be done by changing the absolute cavity length of the interferometer. This was done by changing the offset bias voltage of the interferometer.

8. Demonstration of a cloud of cold atoms

After controlling and locking the laser frequencies and finding the necessary magnetic field strength it was possible to observe a cloud of atom that was visible with the surveillance camera. Simultaneously we observed the cloud with our second CCD camera. The correct magnetic field direction was found by trial and error. For this we changed the polarity on the magnetic field power supplies while adjusting the best laser frequencies. Fig.26 shows a typical image obtained with our surveillance camera in our initial setup.

Fig. 26. Cloud of atoms obtained with surveillance camera. Left: no cloud, right: cloud of cold atoms.

Fig. 27 shows image taken with a modified Samsung photo camera. In this case we removed the optics from the camera and the IR filter. We placed a 50 mm camera lens in front of the camera.

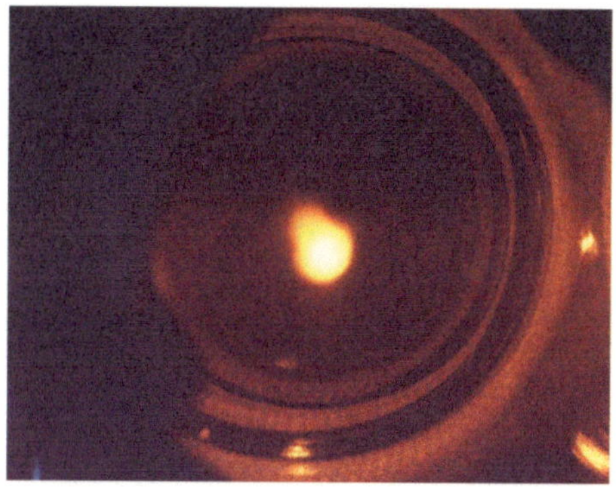

Fig. 27. Image taken with modified Samsung camera. The chamber can be seen.

Fig. 28 shows the cloud image obtained with the IR Altec Vision CCD camera. The cloud diameter was nearly 2.0 mm at its full width.

Fig. 28. Cloud of atoms obtained with our Altec Vision IR CCD camera. Left is the cloud image and right a 3D plot of intensity of the same cloud.

8.1 Optical method: using a Glan Thomson polarizer for laser intensity imbalance

We introduced an optical method, previously developed for laser printers (Duarte, 2005), to study the effect of the laser intensity imbalance on the cloud formation. The method uses two Glan Thomson polarizers to produce a controlled imbalance between pump and probe laser. Each Glan Thomson polarizer was installed in front of the polarizing beam splitter cubes the produces the first division of the cooling and repumping laser respectively as seen in Fig.13. The laser intensity was controlled at will by rotating the Glan Thomson polarizer. The polarizing beam splitter cube contributes for further reduction of the laser intensity. We measured the intensity behind the beam splitter cube after each intensity reduction and recorded simultaneously the cloud with our camera. The laser polarization was slightly rotated after passing the beam splitter polarizing cube not affecting the overall functioning of the cloud. A more precise method could be realized by fixing the Glan Thomson polarizers for maximum transmission and rotating at will the field in front ofs each Glan Thomson polarizer by means of a half wave plate disposed in front of it. By this method the laser field polarization would be kept fixed after passing the Glan Thomson.

9. Study of intensity imbalance on cloud formation

The maximum optical power for the cooling and repumping lasers was nearly 48 mW. The imbalance was started keeping the repumping laser at 48 mW and changing the optical power from the cooling laser. The visibility of the cold cloud reached its minimum value as the power was decreased to 10 mW that is nearly 1/5 of its initial value. On the other hand as the cooling laser is kept at 48 mW, the optical power from the repumping laser was decreased to up to 103 microwatts. At this power the cloud was faintly visible. The power ratio between full visibility and threshold was 1/466 for the repumping laser when the cooling laser was kept at its maximum value. In summary the lasers had large intensity difference and the cloud was still visible. To our knowledge this is the largest power difference disclosed in the open literature between the repumping and cooling lasers.

10. Conclusion

We cooled and trapped rubidium atoms in a magneto optical trap and proved the stability of the cloud for different laser intensities. We studied the effect of laser intensity imbalance on cloud formation. We found that the cloud was still visible when the repumping laser intensity was at 1/466 part of its maximum intensity with the cooling laser at its maximum intensity with typical maximum power of 49 mW for each laser. Decreasing the cooling laser intensity to 1/5 of its maximum value produced destruction of the cloud.

11. Acknowledgment

We are grateful to F. J. Duarte for valuable discussions. We also thank Proyecto DICYT 041131OB, Universidad de Santiago de Chile, Usach.

12. References

Demtröder, W. (1995). Laser Spectroscopy: Basic Concepts and Instrumentation, Springer: Berlin

Duarte, F. (2005). Laser sensitometer, US Patent 6 903 824 B2.

Metcalf, H.; van der Stratten, P. (1999). Laser Cooling and Trapping; Springer: Berlin

Milonni, P.; Eberly, J. (2010). Laser Physics, John Wiley and Sons, Inc., ISBN 978-0-470-38711-9, New Jersey

Olivares, I.; (2007). Selective laser excitation in lithium, Optics Journal, Vol. 1, pp. 7-12

Olivares, I.; (2008). Lithium spectroscopy using tunable diode lasers, Tunable Laser Applications, Chapter 11, ed. F. J. Duarte, Marcel Dekker, New York

Olivares, I.; Aguilar, F.; J. G. Aguirre-Gómez. (2008). Cold atoms observed for the first time at the Universidad de Santiago de Chile, Journal of Physics, Conference Series 134

Olivares, I.; Cuadra, J.; Aguilar, F.; Aguirre, J.; Duarte, F. (2009). Optical method using rotating Glan-Thompson polarizers to independently vary the power of the excitation and repumping lasers in laser cooling experiments, Journal of Modern Optics, Vol. 56, pp.1780-1784

Olivares, I.E, Duarte, A.E., Saravia, E.A, Duarte, F. J. ,Lithium isotope separation with tunable diode lasers. *Appl. Optics*, Vol.41 (2002) p.2973-2977.

Rapol, U.; Wasan, A.; and Natarajan V. (2001).Loading of a Rb magneto-optic trap from a getter source, Physical Review A 64, 023402

Wieman, C.; Flowers, G.; Gilbert, S. (1995). Inexpensive laser cooling and trapping experiment for undergraduate laboratories, American Journal of Physics, Vol.63, No.4.; pp. 317-330